P9-AFE-353

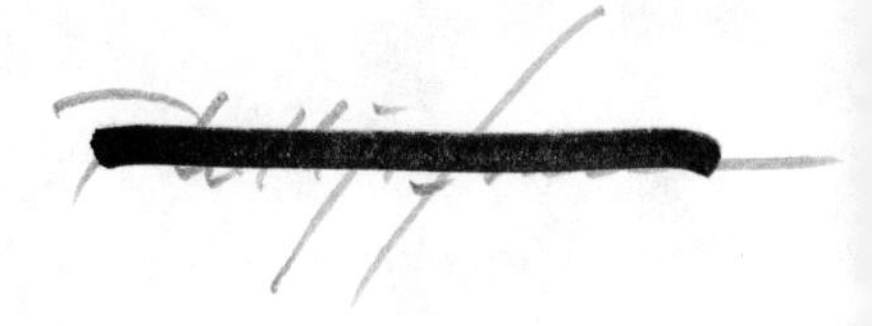

Building Soils for Better Crops

VOLUME 2 IN THE SERIES
Our Sustainable Future

BUILDING SOILS FOR BETTER CROPS

Organic Matter Management

Fred Magdoff

University of Nebraska Press
Lincoln and London

Manufactured in the United States of America

The paper in this book meets the minimum requirements of American National
Standard for Information Sciences – Permanence of
Paper for Printed Library Materials,
ANSI Z39.48–1984.
Library of Congress Cataloging-in-Publication Data
Magdoff, Fred, 1942–
Building soils for better crops : organic matter
management / Fred Magdoff.
p. cm. – (Our sustainable future ; v. 2)
Includes bibliographical references and index.
ISBN 0-8032-3160-1 (cl)
1. Humus. 2. Soil management.
I. Title. II. Series.
S592.8.M34 1993 631.4'17 –
dc20 92-2362
CIP

• • •

To the memory of my brother, Michael.

His short life has been a source

of strength and a reminder of my debt

to those who have gone before.

• • •

Contents

PART TWO: PRACTICES

PART THREE: DYNAMICS AND CHEMISTRY

• • •

Tables and Figures

TABLES

FIGURES

• • •

Preface

To use the land without abusing it . . .
—J. Otis Humphrey, early 1900s

I have written this book with farmers, gardeners, extension agents, and students in mind. My goal is to get people to think about soil organic matter when they plan crop-growing practices. If you follow practices that build up and maintain good levels of soil organic matter, you will find it easier to grow healthy and high-yielding crops. Plants will be able to withstand droughty conditions better, and won't be as bothered by insects and diseases. There will also be less reason to use as much commercial fertilizer and lime as many farmers now purchase. Soil organic matter is that important!

There are a number of previously published books about soil organic matter. Some deal with the basic chemistry of organic molecules in the soil and some are about soil biology. None of these books deal extensively with the details of organic matter management. This book provides an essential introduction to soil organic matter and the practical issues and strategies of soil organic matter management. For those interested in a deeper understanding of the subject, the last two chapters contain a simple theoretical approach to organic matter dynamics, nutrient availability, and the chemistry of organic matter.

A book like this one cannot give exact answers to problems on specific farms. There are just too many differences from one field to another and one farm to another to warrant blanket recommendations. In order to make specific suggestions it is necessary to know the details of the soil, crop, climate, machinery, human considerations, and other variable factors. This book is meant to give the reader an appreciation of the importance of soil organic matter and to suggest practices that build up and maintain this key to soil fertility.

People over many centuries have had an appreciation for the importance of the issues we struggle with today. I quote some of these persons in epigraphs at the beginning of each chapter in appreciation for those who have come before. I was especially fascinated by Vermont Agricultural Experiment Station Bulletin No. 135 published in 1908. It contains an article by three scientists about the importance of soil organic matter that is strikingly modern in many ways. I guess the saying is right—what goes around comes around. Sources cited at the end of chapters are those I referred to during writing. They are not a comprehensive list of references on the subject.

A number of people reviewed the manuscript at one stage or another and made very useful suggestions. I would like to thank the following: Bill Jokela, Vern Grubinger, Wendy Sue Harper, David Stern, Elizabeth Henderson, Rich Bartlett, and Anthony Potenza. I would also like to thank Huck Gutman and my wife, Amy Demarest, for helping to edit the manuscript. Any mistakes are, of course, mine alone.

• • •

Introduction

Used to be anybody could farm. All you needed was a strong back . . . but nowadays you need a good education to understand all the advice you get so you can pick out what'll do you the least harm.—Vermont saying, mid-1900s

Many farmers, agricultural scientists, and extension specialists now feel that there are severe problems associated with conventional agriculture. For example, using too much nitrogen fertilizer or animal manure sometimes causes elevated nitrate concentrations in ground water. These concentrations can become high enough to pose a health hazard. Phosphate in runoff water enters lakes and degrades their waters by stimulating aquatic-weed growth. Antibiotics used to fight diseases in farm animals can enter the food chain and be found in the meat we eat. Erosion associated with conventional tillage and lack of good rotations degrades our precious soil and at the same time causes the silting up of reservoirs, ponds, and lakes. The high cost of purchased items (inputs) used in agriculture has caught farmers in a cost-price squeeze that makes it hard to run a profitable farm.

The food we eat and our surface and ground waters are sometimes contaminated with chemicals used in agriculture. Pesticides used to control weeds, insects, and plant diseases are found in foods, animal feeds, groundwater, and in surface water running off agricultural fields. Farmers and farm workers are at special risk. Recent studies have shown higher cancer rates among those who work with or near certain pesticides.

Given these numerous problems, farmers cannot continue to farm as they have. The system is not sustainable. A major effort is now underway to develop and implement sustainable agricultural practices. Sustainable practices must be both more environmentally sound than conventional practices and at the same time more economically rewarding for farmers. As farmers use management skills and better knowledge to work more closely together with the biological world, they will frequently find that there are ways to lessen purchases of products from off the farm. With a new emphasis on sustainable agriculture has come a reawakening of interest in soil organic matter.

Early scientists, farmers, and gardeners were well aware of the importance of organic matter to the productivity of soil. The significance of soil organic matter, including living organisms in the soil, was understood by scientists at least as far back as the seventeenth century. John Evelyn, writing in England during the 1670s, described the importance of topsoil and explained that the productivity of soils tended to be lost with time but that its fertility could be maintained by adding organic residues. Charles Darwin, the great natural scientist of the nineteenth century who developed the modern theory of evolution, studied and wrote about the importance of earthworms to the recycling of nutrients and the general fertility of the soil. Around the turn of the twentieth century there was again an appreciation of the importance of soil organic matter. Scientists had realized that "worn out" soils, where productivity had drastically declined, resulted mainly from the depletion of soil organic matter. But at the same time, scientists could see a transformation coming: although organic matter was "once extolled as the essential soil ingredient, the bright particular star in the firmament of the plant grower, it fell like Lucifer" under the weight of "modern" agricultural ideas (Hills, Jones, and Cutler, 1908). With the availability of inexpensive fertilizers after World War II, and of cheap water for irrigation in some parts of the western United States, most people working with soils forgot or ignored the importance of organic matter.

A new logic developed that held that most soil-related problems could be dealt with by increasing external inputs. If a soil were deficient in some nutrient, you bought a fertilizer and spread it on the soil. If a soil didn't store enough rainfall, all you needed was irrigation. If a soil became too compacted and water couldn't penetrate into the soil, you could use an implement such as a chisel to tear it open. If a plant disease or insect infestation occurred, you applied a pesticide.

Some of these problems, such as soil fertility, water storage, and compaction, are directly related to soil organic matter. Others, such as disease and insect pest infestations, may be indirectly related to soil organic matter. If we are to work with the biological and physical world instead of attempting to overwhelm and dominate it, the buildup and maintenance of good levels of organic matter in our soils is of critical importance.

This book has three parts. Part I provides background information about organic matter: what it is, why it is so important to soil fertility, and why different soils have different amounts of it. Part II deals with practices that promote soil organic matter buildup and maintenance. These practices include the use of animal manures and cover crops, residue management, rotations, composts, reduced tillage, and erosion control. Following practices that build up and maintain organic matter may be the key to soil fertility and help solve most problems, but other soil-management practices are also needed to monitor and supplement the management of soil organic matter. These practices, such as soil testing and the use of lime and fertilizers, are briefly discussed. Part III covers the chemistry and dynamics of soil organic matter. These chapters are included for those readers who would like a more extensive discussion of the function of soil organic matter in the soil. The practical material in Parts I and II can be read and understood without reading Part III.

Source

Hills, J. L., C. H. Jones, and C. Cutler. 1908. Soil deterioration and soil humus. In Vermont Agricultural Experiment Station Bulletin 135, pp. 142–77. College of Agriculture, Burlington, Vermont.

Part One: The Basics

1

• • •

What Is Soil Organic Matter?

Follow the appropriateness of the season, consider well the nature and conditions of the soil, then and only then least labor will bring best success. Rely on one's own idea and not on the orders of nature, then every effort will be futile.—Jia Si Xie, 6th century, China

Soil consists of four important parts or fractions: solid minerals, water, air, and organic matter. The solid minerals, starting with the largest particle size, are sand, silt, and clay. They mainly consist of silicon, oxygen, aluminum, potassium, calcium, and magnesium. The soil water, also called the soil solution, contains dissolved nutrients and is the main source of water for plants. Essential nutrients are made available to the roots of plants through the soil solution. The air in the soil, in contact with the air aboveground, provides roots with oxygen and helps remove excess carbon dioxide from respiring root cells. Organic matter has an overwhelming effect on almost all soil properties. It consists of three distinctly different parts—living organisms, fresh residues, and well-decomposed residues. These three parts of soil organic matter have been described as the living, the dead, and the very dead. This three-way classification may seem simple and unscientific, but it is very useful.

The living part of soil organic matter includes a wide variety of microorganisms such as bacteria, viruses, fungi, protozoa, and algae. It even includes plant roots and the insects, earthworms, and larger animals such as moles, woodchucks, and rabbits that spend some of their time in the soil. The living portion represents about 15% of the total soil organic matter. Microorganisms, earthworms, and insects help break down crop residues and manures by mixing them with the minerals in the soil and in the process recycle energy and plant nutrients. Sticky substances on the skin of earthworms and those produced by fungi help bind particles together. This helps to stabilize the soil *aggregates,* those clumps of particles that make up good soil structure. Organisms such as earthworms and some fungi also help directly to stabilize the soil's structure (for example, by producing channels that allow water to infiltrate) and thereby improve soil water and aeration status. A good soil structure reduces water runoff from the soil and lessens erosion. Plants roots also interact in significant ways with the various microorganisms and animals living in the soil. Further discussion of the interactions between soil organisms and roots is provided in Chapter 2.

The fresh residues, or "dead" organic matter, consist of recently deceased microorganisms, insects, earthworms, old plant roots, crop residues, and recently added manures. In some cases just looking at them is enough to identify what the fresh residues have been. Through decomposition, these materials become food for living microorganisms, insects, and earthworms. As organic materials are decomposed, many nutrients needed by plants are released. Chemicals produced during the decomposition of fresh residues also help to bind soil particles together and give the soil a good structure.

Organic molecules directly released from cells of fresh residues, such as proteins, amino acids, sugars, and starches, are also considered part of this fresh organic matter. These molecules generally do not last long in the soil because there are so many microorganisms that use them. This part of soil organic matter is the active, or easily decomposed, fraction. This active fraction of soil organic matter is the main supply of food for various organisms living in the soil.

The well-decomposed organic material in soil, the "very dead," is called *humus*. Humus is a term sometimes used to describe all soil organic matter. Some use it to describe just the part you can't see without a microscope. I will use the term to refer only to the well-decomposed part of soil organic matter. The already well decomposed humus is not a food for organisms, but its very small size and chemical properties make it a very important part of the soil. Humus holds on to some essential nutrients, storing them for slow release to plants. Humus also can surround certain potentially harmful chemicals and prevent them from causing damage to plants. Good amounts of soil humus can both lessen drainage or compaction problems that occur in heavy, clay soils and improve water retention in light, sandy soils.

Organic matter decomposition is a process that is similar to the burning of wood in a stove. When burning wood reaches a certain temperature, the carbon in the wood combines with oxygen from the air and forms carbon dioxide. As this occurs, the energy stored in the carbon-containing chemicals in the wood is released as heat in a process called oxidation. The biological world, including humans, animals, and microorganisms, also makes use of energy inside carbon-containing molecules. This process of converting sugars, starches, and other compounds into a directly usable form of energy is also a type of oxidation. We usually call it *respiration.* Oxygen is used and carbon dioxide and heat are given off in this process.

A multitude of microorganisms, earthworms, and insects get their energy and nutrients by decomposing organic residues in soils. At the same time, much of the energy stored in residues is used by organisms to make new chemicals as well as new cells. How does energy get stored inside organic residues in the first place? Green plants use the energy of sunlight to link carbon atoms together into larger molecules. This process, known as *photosynthesis,* is used by plants to store energy for respiration and growth.

Summary

Soil organic matter consists of a large number of organisms living in the soil, fresh residues of various kinds, and the well-decomposed material called humus. Organic materials decompose in soils as organisms use the residues as sources of energy and nutrients.

Source

Brady, N. C. 1990. *The nature and properties of soils.* 10th ed. New York: Macmillan Publishing Co.

2

• • •

The Living Soil

The plough is one of the most ancient and most valuable of man's inventions; but long before he existed the land was in fact regularly ploughed, and still continues to be thus ploughed by earthworms. It may be doubted whether there are many other animals which have played so important a part in the history of the world, as have these lowly organized creatures.—Charles Darwin, 1881

Soil organisms are similar to plants and animals in many ways. When soil organisms and roots go about their normal functions of getting energy for growth from organic molecules they "respire" and give off carbon dioxide to the atmosphere. A soil system can also get sick in the sense that it may become incapable of supporting healthy plants. The organisms, both large and small, that live in the soil play a significant role in maintaining a healthy soil system and healthy plants.

We are mainly interested in the organisms living in the soil because of their role in decomposing organic matter and incorporating it into the soil. As organic matter is decomposed, nutrients are made available to plants, humus is produced, soil aggregates are formed, chan-

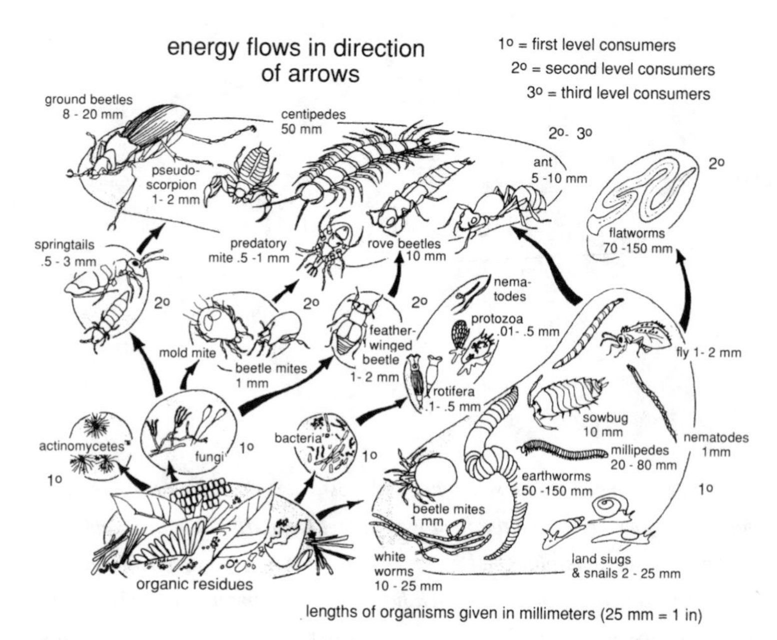

Figure 2.1. Soil organisms and their role in decomposing residues. From D. L. Dindal, Soil organisms and stabilizing wastes, *Compost Science/Land Utilization* (July/August 1978), p. 9. Used with permission of the JG Press.

nels are created for water infiltration and better aeration, and organic matter originally on the surface is brought deeper into the soil. Soil organisms influence every aspect of decomposition and nutrient availability.

There are several different ways to classify soil organisms. For example, each type of organism can be discussed separately or all organisms that do the same types of things can be discussed together. We can also discuss organisms by general size, such as very small, small, medium, large, and very large. Or we can look at soil organisms according to their role in the decomposition of organic materials. Organisms that use fresh residues as their source of food are called primary, or first-level, consumers of organic materials (see figure 2.1). These primary consumers break down large pieces of residues into smaller fragments. Secondary consumers are organisms that feed on the primary consumers themselves or their waste products. Tertiary consumers then feed on the secondary consumers.

There is constant interaction among the organisms living in the soil. Some organisms help other organisms, as when bacteria that live inside the earthworm's digestive system help decompose food. Organisms may also directly compete with each other for the same food. Some organisms naturally feed on others. To say "it's a jungle down there" is a good way of looking at the soil environment, since the interrelationships of soil organisms are as complex as those that occur aboveground in a rain forest.

Some soil organisms can harm plants either by causing disease or by being parasites. In other words, there are "good" as well as "bad" bacteria, fungi, nematodes, and insects. One of the goals of agricultural production should be to create conditions that enhance the growth of beneficial organisms while decreasing populations of those that are potentially harmful.

Soil Microorganisms

Microorganisms are very small forms of life that can usually live as single cells, although many also form colonies of cells. A microscope is usually needed to see individual cells of these organisms. There are usually many more microorganisms in topsoil, where food sources are plentiful, than in subsoil. These organisms are important primary decomposers of organic matter, but they do other things as well to help growing plants.

Bacteria

Bacteria can live in almost any habitat. They are found inside the digestive system of animals, in the ocean and fresh water, in compost piles (even at temperatures over 130°F), and in soils. They are very plentiful in soils: a single gram of topsoil commonly contains over 10 million bacteria! A teaspoon of topsoil may contain over 50 million bacteria. Although some kinds of bacteria can live in flooded soils without oxygen, most require well-aerated soils. In general, bacteria tend to do better in neutral soils than in acid soils.

In addition to being some of the first organisms to begin decomposing residues in the soil, many bacteria help plants by increasing nutrient availability. For example, there are large numbers of bacteria that can dissolve phosphorus, making it more available for plants to use.

Bacteria are also very helpful in providing nitrogen to plants. Although nitrogen is needed in large amounts by plants, it is commonly deficient in agricultural soils. You may wonder how soils can be deficient in nitrogen when we are surrounded by it: 78% of the air we breathe is composed of nitrogen gas. But plants as well as animals face the dilemma of the Ancient Mariner, who was adrift at sea without fresh water: "water, water, everywhere but not a drop to drink." Unfortunately, neither animals nor plants can use nitrogen gas (N_2) for their nutrition. However, some types of bacteria are able to take nitrogen gas from the atmosphere and convert it into a form that plants can use to make amino acids and proteins. This conversion process is known as nitrogen fixation.

Some nitrogen-fixing bacteria form symbiotic, or mutually beneficial, relationships with plants. One such symbiotic relationship that is very important to agriculture is the nitrogen-fixing rhizobia group of bacteria that live in nodules formed on the roots of legumes. These bacteria provide nitrogen in a form that leguminous plants can use while the legume provides the bacteria with sugars for energy.

People eat some legumes or their products, such as peas, dry beans, and tofu made from soybeans. Soybeans, alfalfa, and clover are used for animal feed. Clovers and hairy vetch are grown as *cover crops* to enrich the soil with organic matter as well as available nitrogen. In an alfalfa field, the bacteria may fix hundreds of pounds of nitrogen per acre each year. With peas, the amount of nitrogen fixed is much lower, around 30 to 50 pounds per acre.

Lignin is a large molecule found in mature plant tissue that is difficult for most organisms to break down. Lignin also frequently protects other molecules like cellulose from decomposition. The *actinomycetes,* another group of bacteria, break large lignin molecules into smaller sizes. Actinomycetes have some characteristics similar to fungi but are sometimes grouped by themselves and given equal billing with bacteria and fungi.

FUNGI

Fungi are another important type of soil microorganism. Yeast is a fungus used in baking and in the production of alcohol; a number of antibiotics are produced by other fungi. We have all probably let a loaf of bread sit around too long only to find fungus growing on it. And we have seen or eaten mushrooms, the fruiting structure of some fungi. Farmers know that plant diseases such as downy mildew, damping-off, various types of root rot, and apple scab are caused by fungi. Fungi are also important in starting the decomposition of fresh organic residues. They help get things going by softening organic debris and making it easier for other organisms to join in the decomposition process. Fungi are less sensitive to acid-soil conditions than are bacteria. None are able to function without oxygen.

Almost all plants develop a beneficial relationship with fungi that increases the contact of roots with the soil. Fungi infect the roots and send out rootlike structures called *hyphae.* The hyphae take up water and nutrients that can then feed the plant. This is especially important for phosphorus nutrition of plants in low-phosphorus soils. The hyphae help the plant absorb water and nutrients and in return the fungi receive energy in the form of sugars, which the plant produces in its leaves and sends down to the roots. This symbiotic interdependency between fungi and roots is called a mycorrhizal relationship. All things considered, it's a pretty good deal for both the plant and the fungus.

Algae

Algae, like crop plants, convert sunlight into energy. They are found in abundance in the flooded soils of swamps and rice paddies. They can also be found on the surface of poorly drained soils or in wet depressions. Some algae form mutually beneficial relationships with other organisms. Lichens found on rocks are an association between a fungus and an alga.

Protozoa

Protozoa are single-celled animals that use a variety of means to move about in the soil. Like bacteria and many fungi, they can be seen only with the help of a microscope. They are mainly secondary consumers of organic materials, feeding on bacteria, fungi, other protozoa, and organic molecules dissolved in the soil water.

Small and Medium-sized Soil Animals

Nematodes

Nematodes are simple soil animals that resemble small worms. They tend to live in the water films around soil aggregates. Some types of nematodes feed on plant roots and are well known as plant pests. Diseases such as pythium and fusarium that enter feeding wounds on the root sometimes cause more damage than the feeding itself. Many other types of nematodes help in the breakdown of organic residues and feed on fungi, bacteria, and protozoa as secondary consumers.

Earthworms

Earthworms are every bit as important as Charles Darwin believed more than a century ago. They are keepers and restorers of soil fertility. There are different types of earthworms, including the nightcrawler, field (garden) worm, and manure (red) worm. Earthworms have different feeding habits. Some are primary consumers, feeding on plant residues that remain on the soil surface, while other types tend to feed on organic matter that is already mixed with the soil.

The surface-feeding nightcrawlers fragment and mix fresh residues with minerals, bacteria, and enzymes in their digestive system and the resulting material is given off as worm casts. They also bring food down into their burrows and thereby mix organic matter deep into the soil. Earthworms feeding on debris already below the surface continue to decompose organic materials and mix them with minerals. In general, worm casts are higher in available plant nutrients such as nitrogen, calcium, magnesium, and phosphorus than the surrounding soil and therefore make an important contribution to the nutrient needs of plants.

A number of types of earthworms, including the surface-feeding nightcrawler, make burrows that allow rainfall to infiltrate into the soil easily. These worms usually burrow to 3 feet or more under dry conditions. Even those types of worms that don't normally produce channels to the surface do help loosen the soil and produce channels and cracks below the surface that help water and air to move.

Earthworms do some unbelievable work. They move a lot of soil from below up to the surface—from about 1 to 100 tons per acre each year. One acre of soil 6 inches deep weighs about 2 million pounds, or 1,000 tons. So 1 to 100 tons is the equivalent to about 6 thousandths of an inch to about half an inch of soil. The number of earthworms in the soil ranges from close to zero to hundreds of thousands and even over a few million per acre. Just imagine, if you create the proper conditions for earthworms you can have 800,000 small channels per acre that conduct water into your soil during downpours.

Earthworms do best in well-aerated soils that are supplied with plentiful amounts of organic matter. A study in Georgia showed that soils with higher amounts of organic matter contained higher numbers of earthworms. Surface feeders, a type we would especially like to encourage, need residues left on the surface. They are harmed by plowing, which disturbs their burrows and buries their food supplies. Worms are usually more plentiful under no-till practices than when conventional tillage systems are used. Although most pesticides have little effect on worms, others, such as aldicarb, parathion, and heptachlor, are very harmful to earthworms.

Diseases or insects that overwinter on leaves of crops can sometimes be partially controlled by high earthworm populations. Worms eat the leaves and incorporate the residues deeper into the soil. The apple scab fungus, a major pest of apples in humid regions, can be partly controlled in this way, as can some leaf miner insects.

If you don't have enough earthworms, it's possible to increase their numbers by bringing in worms. Work is progressing well on experiments to add earthworm cocoons, containing eggs, to soil. Once this system is perfected it may become economically worthwhile to introduce large numbers of worms into a field. But if favorable conditions for worms aren't present in your soil, their introduction may be a waste of time and money.

Insects

Insects are another group of animals that may inhabit soils. Common types of soil insects include termites, ants, fly larvae, and beetles. Many insects are secondary and tertiary consumers. Many beetles, in particular, eat other types of soil animals. Termites, well-known feeders on woody material, also feed on decomposed organic residues in the soil.

Other small to medium-sized soil animals include millipedes, centipedes, mites, springtails, slugs, snails, and spiders. Millipedes are primary consumers of plant residues while centipedes tend to feed on other organisms. Mites may feed on such food sources as fungi, other mites, and insect eggs, although some feed directly on residues. Springtails feed on fungi and animal remains. Spiders feed mainly on insects and their role in keeping insect pests from developing large populations can be important.

Large Animals

Large animals such as moles, rabbits, woodchucks, snakes, prairie dogs, and badgers burrow in the soil and spend at least some their lives belowground. Moles are secondary consumers, with their diet consisting mainly of earthworms. Most of the other animals exist on vegetation. In many cases their presence is considered a nuisance in agricultural production or in lawns and gardens. But their burrows may help conduct water away from the surface during downpours and thus decrease erosion.

Plant Roots

Healthy plant roots are essential for good crop yields. Roots are clearly influenced by the soil in which they live. If the soil is compact, low in nutrients or water, or has other problems, plants will not grow well. But plants also influence the soil in which they grow. The physical pressure of roots growing through the soil helps form aggregates by bringing particles closer together. Small roots also help bind parti-

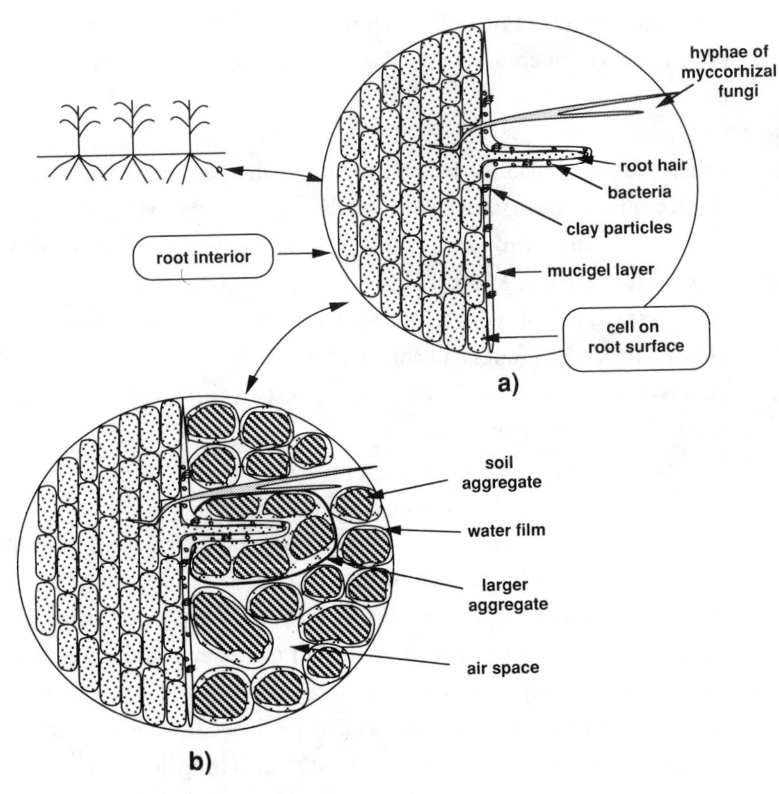

Figure 2.2. Closeup view of a plant root. a) The mucigel layer containing some bacteria and clay particles on the outside of the root. Also shown is a myccorhizal fungus sending out its rootlike hyphae into the soil. b) Soil aggregates surrounded by thin films of water. Plant roots take water and nutrients from these films. Also shown is a larger aggregate made up of smaller aggregates pressed together and held in place by the root and hyphae.

cles together. In addition, many organic compounds are given off, or exuded, by plant roots and provide nourishment for soil organisms living on or near the roots. There is also a sticky layer surrounding roots, called the *mucigel,* that provides for very close contact between microorganisms, soil minerals, and the plant.

For plants with extensive root systems, such as the grasses, the amount of living tissue belowground may actually weigh more than the amount of leaves and stems we see aboveground.

Biological Diversity

A diverse biological community in soils is important in maintaining a healthy environment for plants. There are numerous soil bacteria and fungi that can cause plant diseases. There are insects and parasitic nematodes that can be major pest problems for crops. Diverse populations of soil organisms maintain a system of checks and balances that can keep disease organisms or parasites from becoming major plant problems. There are fungi that kill nematodes and others that kill insects. Some fungi produce antibiotics that kill bacteria. Protozoa feed on bacteria. There are also bacteria that kill harmful insects.

Certain farming practices increase biological diversity in the soil. For example, organic residues that are regularly returned to the soil provide a food supply for a diverse group of organisms. Organic materials may include crop residues, green manures, animal manures, and composts. Leaves and grass clippings may be an important source of organic residues for gardeners. Crop rotation is another way to help maintain a healthier biological mix. When crops are rotated regularly there are fewer parasite, disease, weed, and insect problems than when the same crop is grown year after year. Frequent cultivation reduces the number of soil organisms, so decreasing tillage will also help increase the number of organisms present in soil.

Summary

The huge number of organisms living in the soil is responsible for the decomposition of organic matter. The interrelationships of organisms are complex, with competition for food and feeding of one on another. A continuous supply of residues and manures creates a healthy mix of organisms, maintaining a biological system of checks and balances. Plant roots are also part of the living portion of soil organic matter. Microorganisms form associations with plant roots, many colonizing the zone around roots.

Sources

Alexander, M. 1977. *Introduction to soil microbiology.* 2d ed. New York: John Wiley & Sons.

Hendrix, P. F., M. H. Beare, W. X. Cheng, D. C. Coleman, D. A. Crossley, Jr., and R. R. Bruce. 1990. Earthworm effects on soil organic matter dynamics in aggrading and degrading agroecosystems on the Georgia Piedmont. *Agronomy Abstracts,* p. 250. This is the reference for the Georgia study on earthworms.

Paul, E. A., and F. E. Clark. 1989. *Soil microbiology and biochemistry.* San Diego: Academic Press.

3

• • •

Why Is Soil Organic Matter So Important?

Why are soils which in our father's hands were productive now relatively impoverished?—J. L. Hills, C. H. Jones, and C. Cutler, 1908

A fertile soil is the foundation for healthy plants, animals, and humans. Soil organic matter is the foundation for productive soils. Understanding the role of organic matter in maintaining a healthy soil is essential to developing sustainable agricultural practices. While it's true that plants can be grown in gravel or sand hydroponic systems without soil, large-scale hydroponic systems are usually neither economically nor ecologically sound. These systems require intensive management and careful control of nutrient levels. Although they are sometimes used for commercial vegetable production, especially lettuce, high start-up and operating costs frequently make them unprofitable.

As soil organic matter decreases, it becomes increasingly difficult to grow plants, because fertility, water availability, compaction, erosion, parasites, disease, and insect problems become more common. Ever higher levels of inputs—fertilizers, irrigation water, pesticides, and machinery—are required to maintain yields in the face of organic matter depletion. But, if attention is paid to proper organic matter management, the soil can support a good crop without the need for expensive fixes.

The organic matter content of agricultural topsoil is usually in the range of 1 to 6%. A study of soils in Michigan demonstrated potential crop-yield increases of about 12% for every 1% organic matter. You might wonder how something that's only a small part of the soil can have such importance to growing healthy and high-yielding crops. Part of the explanation for this influence is the small particle size of the well-decomposed portion of organic matter—the humus. Its large surface area to volume ratio means that humus is in contact with a considerable portion of the soil. The intimate contact of humus with the rest of the soil allows many reactions such as the release of available nutrients into the soil water, to occur rapidly.

Plant Nutrition

Plants need seventeen chemical elements for their growth—carbon (C), hydrogen (H), oxygen (O), nitrogen (N), phosphorus (P), potassium (K), sulfur (S), calcium (Ca), magnesium (Mg), iron (Fe), manganese (Mn), boron (B), zinc (Zn), molybdenum (Mo), copper (Cu), cobalt (Co), and chlorine (Cl). Plants obtain carbon as carbon dioxide (CO_2) and oxygen partially as oxygen gas (O_2) from the air. The remaining essential elements are obtained mainly from the soil. The availability of these nutrients is influenced either directly or indirectly by the presence of organic matter.

Nutrients from Decomposing Organic Matter

Most of the nutrients in soil organic matter can't be used by plants as long as they exist as part of large organic molecules. But as soil organisms decompose organic matter, nutrients are converted into simpler, inorganic, or mineral, forms that plants can easily use. This process, called *mineralization,* provides much of the nitrogen that plants need by converting it from organic forms such as proteins to ammonium (NH_4^+) and then to nitrate (NO_3^-). Most plants will take up the majority of their nitrogen from soils in the form of the inorganic nitrate molecule. The mineralization of organic matter is also an important mechanism for supplying plants with nutrients such as phosphorus and sulfur and most of the micronutrients, or trace nutrients, which are needed in very small quantities by plants. This release of nutrients from organic matter by mineralization is part of a larger agricultural nutrient cycle (see figure 3.1). Although we are focusing on organic matter, keep in mind that nutrients such as calcium and potassium are also released in available forms when minerals dissolve in the soil.

Storage of Nutrients on Soil Organic Matter

Decomposing organic matter can feed plants directly, but it can also benefit the plant indirectly. Humus has many negative charges. Because opposite charges attract, humus is able to hold onto positively charged nutrients (cations), such as calcium (Ca^{++}), potassium (K^+), and magnesium (Mg^{++}), keeping them from leaching deep into the subsoil when water moves through the topsoil (see figure 3.2a). Nutrients held in this way can be gradually released into the soil solution and made available to plants throughout the growing season.

Clay particles also have negative charges on their surfaces (figure 3.2b). But for light-textured soils low in clay content, organic matter may be the major source of this capacity to hold onto cations. Some types of clays, such as those found in the southeastern United States and in the tropics, tend to have low amounts of negative charge, and

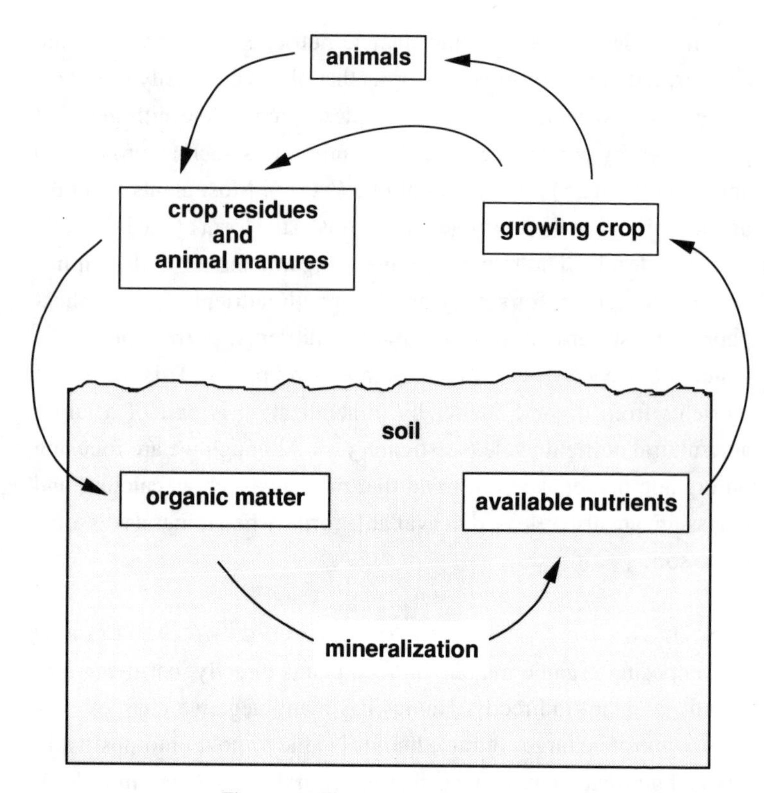

Figure 3.1. The cycle of plant nutrients.

for these soils, as well, organic matter may be the major source of negative charge. The ability of organic matter to hold onto cations in a way that keeps them available to plants is known as *cation exchange capacity* (CEC) and will be discussed in more detail in Chapter 14.

Protection by Chelation

Organic molecules in the soil may also act as *chelates,* holding onto and protecting certain nutrients. In general, elements are held more

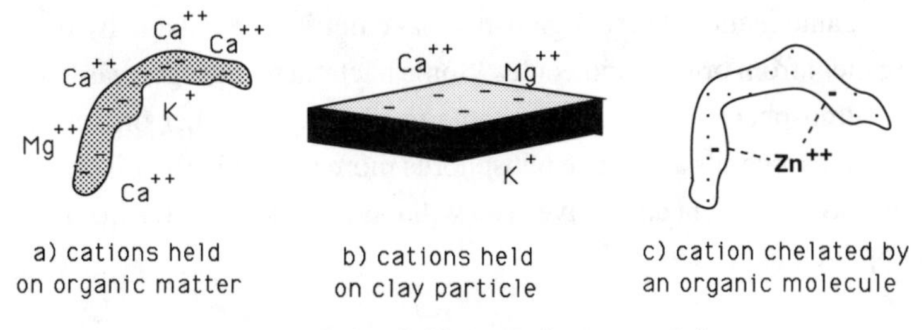

Figure 3.2. Cations held on organic matter and clay.

strongly by chelating molecules than by cation exchange capacity. Chelates work so well because they hold each nutrient at more than one location on the organic molecule (figure 3.2c). In some soils, trace elements such as iron, zinc, and manganese would be converted to unavailable forms if they were not chelated. It is not uncommon to find low organic matter soils or exposed subsoils deficient in available micronutrients.

Other Forms of Protection

There is some evidence that organic matter in the soil can inhibit the conversion of available phosphorus to forms that are unavailable to plants. One explanation is that organic matter coats the surfaces of minerals that can bond tightly to phosphorus. Once these surfaces are covered, available forms of phosphorus are less likely to react with them. In addition, humic substances may chelate aluminum and iron, both elements that can react with phosphorus in the soil solution. When they are held as chelates, these metals are unable to form an insoluble mineral with phosphorus.

Beneficial Effects of Soil Organisms

Soil organisms are essential for keeping plants well supplied with nutrients. The presence of soil organisms is necessary for the breakdown of organic matter. These organisms make nutrients available by freeing them from organic molecules. Some bacteria fix nitrogen gas from the atmosphere, making it available to plants, and other organisms dissolve minerals and make phosphorus more available. If soil organisms aren't present and active, you will have to add more fertilizers to supply plant nutrients.

Soil Tilth

The arrangement and collection of minerals as aggregates and the degree of soil compaction have huge effects on plant growth. When soil has a favorable physical condition for growing plants it is said to have good *tilth*. A soil with good tilth is porous and allows water to enter easily instead of running off the surface. More water is stored in the soil for plants to use between rains and less soil erosion occurs. But good tilth also means that the soil is well aerated, and roots can easily get oxygen and get rid of carbon dioxide. A porous soil does not restrict root development and exploration. As a soil's structure deteriorates and aggregates are broken down, compaction increases and aeration and water storage decrease. A soil layer can become so compacted that roots can't even grow into it.

A soil with excellent physical properties will have numerous channels and pores of many different sizes. The smallest capillary-size pores hold water so tightly that plants are not able to use it. But medium to large capillary-size pores, about 1 to 6 thousandths of an inch in diameter, can hold water against the downward pull of gravity and at the same time make it available to plants. Large channels and cracks in a soil—those that you can easily see—serve to conduct water into the soil under very intense rainfall. They drain quickly after a rainstorm is over. Clay soils low in organic matter may have many very small and medium-size pores but not enough large ones. Moreover, soils containing lots of clay are also fairly sticky and their structure is broken down easily if the soil is worked when wet. Coarse, sandy soils that are low in soil organic matter may have substantial numbers of larger pores, but the pores for storage of available water may not be plentiful. In both of these situations a plentiful supply of organic matter will make the soil more fertile. Studies on both undisturbed and agricultural soils show that as organic matter increases, soils tend to be less compact and have more space for air passage and water storage.

During the decomposition of residues, sticky substances are produced that bind mineral particles together into clumps, or aggregates. Plant roots and the hyphae of fungi also help bind particles together. The development of aggregates is desirable in all types of soils since it promotes better drainage, aeration, and water storage.

Organic matter, as residue on the soil surface or as a binding agent for aggregates near the surface, plays an important role in lessening soil erosion. Surface residues intercept raindrops and decrease their potential to detach soil particles. Surface residues also slow down water as it flows across the soil, giving it a better chance to infiltrate into the soil. The better the development of aggregates and large channels, the greater the ability of soil to conduct water from the surface into the subsoil. Since erosion tends to remove the most fertile part of the soil, it can cause a significant reduction in crop yields. In some soils, the loss of just a few inches of topsoil may result in a yield reduction of 50%. The surface of some soils low in organic matter may seal over, or crust, as rainfall breaks down aggregates, and pores near the surface fill with solids. When this happens, water that can't infiltrate into the soil runs off the field, carrying valuable topsoil particles with it (figure 3.3).

Large soil pores, or channels, are very important because of their

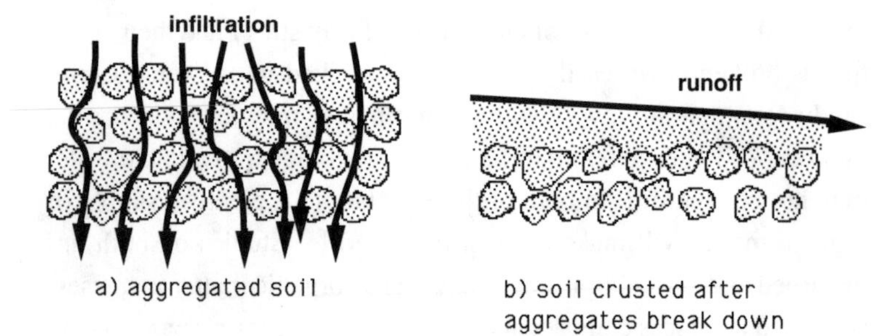

Figure 3.3. Changes in soil surface and water-flow pattern when crusts develop.

ability to allow a lot of water to flow rapidly into the soil. Old root channels may remain open for some time after the root decomposes. Larger soil organisms such as insects and earthworms create channels as they move through the soil. The mucus that earthworms secrete to keep their skin from drying out helps to keep their channels open for a long time. These channels conduct water into the soil during intense rainstorms.

Many times you can tell that one soil is better than another by just looking at them, or picking up some of each, or by how they work up when tilled, or even by sensing how they feel when walked on. For an example of how differences in soil can be created by different management strategies, see the photo of samples from two experiments at the Rodale Research Center in Pennsylvania (figure 3.4). Farmers and gardeners would certainly rather grow their crops on the more porous soil on the left.

Protection of the Soil from Rapid Changes in Acidity

Acids and bases are released as minerals dissolve and organisms go about their normal functions of decomposing organic materials or fixing nitrogen. Acids or bases can also be excreted by the roots of

Figure 3.4. Soil taken from two treatments of an experiment at the Rodale Research Center in Pennsylvania. The soil on the left is from a corn plot in which a cover crop of red clover had been plowed down three months earlier. The soil on the right is from a conventionally managed cornfield (in rotation with soybeans) on which commercial nitrogen fertilizer is used and without a cover crop rotation. Note the more porous soil structure of the soil on the left. Photograph by Dr. G. Bergstrom, used with permission.

plants. Acids also form in the soil from the use of most of the common forms of nitrogen fertilizers. It is best for plants if the soil acidity status, referred to as *pH*, does not swing too wildly during the season. The pH scale is a way of expressing the amount of free hydrogen (H^+) in the soil water. More acid conditions, with greater amounts of hydrogen, are indicated by lower numbers. A soil at pH 4 is very acid. Its solution is 10 times more acid than a soil at pH 5. A soil at pH 7 is neutral—there is just as much base in the water as there is acid. Most crops do best when the soil is slightly acid and the pH is around 6 to 7.

Almost all essential nutrients are more available to plants in this pH range than when soils are either more acid or more basic. Soil organic matter is able to slow down, or *buffer,* changes in pH by taking free hydrogen out of solution as acids are produced or by giving off hydrogen as bases are produced. Soil pH buffering is discussed in more detail in Chapter 14.

Stimulation of Root Development

Humus has a directly beneficial effect on plant growth. While the reasons for this stimulation are not yet understood, certain types of humus cause roots to grow larger and have more branches and this results in larger and healthier plants. Some scientists believe that humic substances act as plant hormones.

Darkening of the Soil

Organic matter tends to darken soils. You can see this easily in coarse-textured sandy soils containing light-colored minerals. Under cool, northern conditions a darker soil surface allows a soil to warm up a little faster in the spring. This provides a slight advantage in seed germination and the early stages of seedling development.

Protection against Harmful Chemicals

There are some naturally occurring chemicals in soils that can harm plants. For example, aluminum is an important part of many soil minerals, and as such poses no threat to plants. But as soils become more acid, especially at pH levels below 5.0 to 5.5, aluminum becomes soluble. Some soluble forms of aluminum, if present in the soil solution, are toxic to plant roots. However, in the presence of significant quantities of soil organic matter the aluminum can be held very tightly and will not do as much damage. Organic matter can also hold onto a number of pesticides allowing them to be detoxified by microbes. Microorganisms can change the chemical structure of some pesticides, industrial oils, and other potentially toxic chemicals, rendering them harmless.

Biological Diversity in Soil

A varied community of organisms is your best insurance against major pest outbreaks and soil fertility problems. A soil rich in organic matter and continually supplied with different types of fresh residues will be home to a more diverse group of organisms than soil depleted of organic matter.

Natural Cycles

The Carbon Cycle

Soil organic matter plays a significant part in a number of global cycles. People are now interested in the carbon cycle because the buildup of carbon dioxide in the atmosphere is thought to cause global warming. A simple version of the carbon cycle, showing soil organic matter's role, is given in figure 3.5. Carbon dioxide is taken up from the atmosphere by plants and used to make all the organic molecules necessary for their lives. Sunlight provides plants with the energy they need to carry out this process. Plants, as well as the animals feeding on plants, release carbon dioxide back into the atmosphere as they use organic molecules for energy in a process called respiration. Carbon dioxide is also released to the atmosphere when fuels such as gas, oil, and wood are burned.

Something that is rarely mentioned when people discuss the carbon cycle is that by far the largest amount of carbon present on the land is not in the living plants, but in soil organic matter. It is estimated that soil organic matter contains about four times the amount of carbon present in living plants. As soil organic matter is depleted it becomes

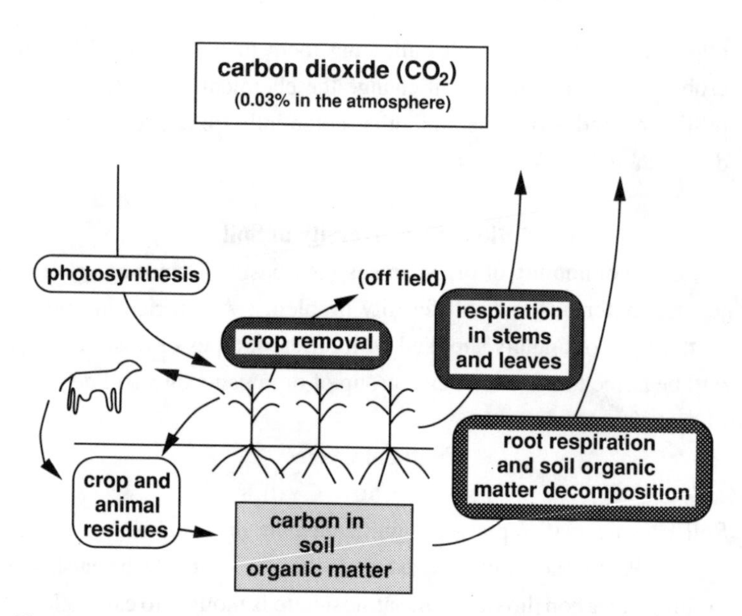

Figure 3.5. The role of soil organic matter in the carbon cycle. Losses of carbon from the field are indicated by a dark border around the words describing the process.

a source of carbon dioxide for the atmosphere. When forests are cleared and burned there is a large CO_2 release to the atmosphere. But there is a potentially larger release of carbon dioxide following conversion of forests to agricultural practices that rapidly deplete the soil of its organic matter.

The Hydrologic Cycle

Organic matter plays an important part in the local, regional, and global water, or hydrologic, cycle due to its role in promoting water infiltration into soils and storage within the soil. Water evaporates from the soil surface and from living plant leaves as well as from the ocean and lakes. Water then returns to the earth, usually far away from where it evaporated, as rain and snow. Soils high in organic matter, with excellent tilth, tend to promote the rapid infiltration of rainwater into the soil. This water may be available for plants to use or it may percolate deep into the subsoil and help to recharge the groundwater supply. Since groundwater is commonly used as a water source for homes in rural areas and for irrigation, recharging groundwater is important. But as the soil's organic matter level is depleted it is less able to accept water and high levels of runoff and erosion usually result. This results in less water for plants and decreased groundwater recharge.

THE NITROGEN CYCLE

Another important global cycle in which organic matter plays a major role is the nitrogen cycle. This cycle is of direct importance in agriculture because available nitrogen for plants is commonly deficient in soils. Figure 3.6 shows a simple version of the nitrogen cycle and how soil organic matter, living and dead, enters into the cycle. Some bacteria living in soils are able to "fix" nitrogen, converting nitrogen gas to forms that other organisms, including crop plants, can use. Something not shown in figure 3.6 is that inorganic forms of nitrogen, ammonium and nitrate, exist in the atmosphere naturally, although air pollution causes higher amounts than normal. Rainfall and snow will then bring inorganic nitrogen forms to the soil. Inorganic nitrogen may also be added in the form of commercial nitrogen fertilizers. These fertilizers are derived from nitrogen gas in the atmosphere through an industrial fixation process.

Almost all of the nitrogen in soils exists as part of the organic matter. But plants are not able to use the nitrogen in organic molecules as their main nitrogen source. Many bacteria and fungi can convert the organic forms of nitrogen into ammonium and a few bacteria can

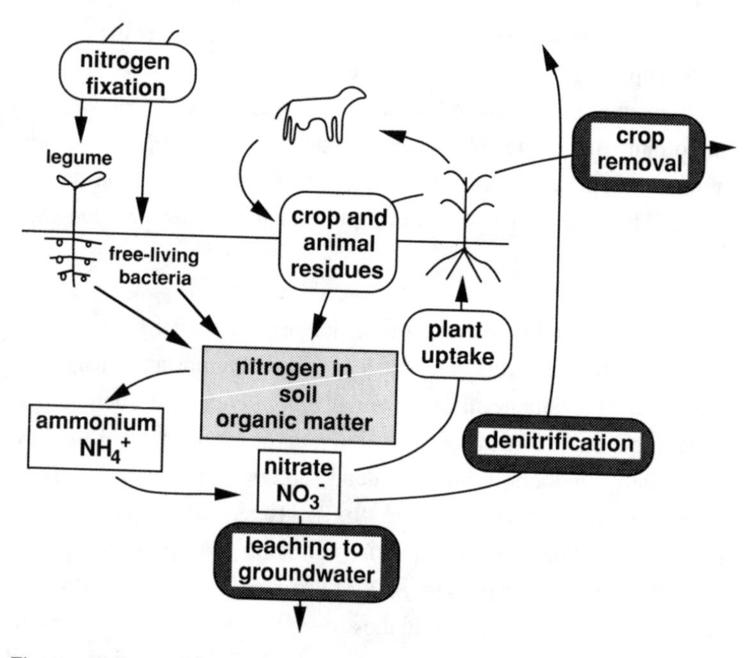

Figure 3.6. The role of soil organic matter in the nitrogen cycle. Losses of nitrogen from the field are indicated by a dark border around the words describing the process.

convert ammonium into nitrate. Both nitrate and ammonium can be used by plants.

Nitrogen can be lost from a soil in a number of ways. When crops are removed from fields, nitrogen and other nutrients are also removed. The nitrate form of nitrogen leaches readily from soils and may end up in groundwater in higher concentrations than may be good for drinking. Nitrate and ammonium may be lost with water runoff (not shown in the figure). Nitrogen once freed from soil organic matter may be converted to forms that end up back in the atmosphere. Bacteria may also convert nitrate (NO_3^-) to nitrogen (N_2) and nitrous oxide (N_2O) gases in a process called denitrification.

Summary

As the organic matter content of soil decreases, its general fertility also decreases as nutrients are less available, soil structure deteriorates, and harmful substances do more damage. Plants that are weakened by poor nutrition, poor aeration, or low water availability are more susceptible to diseases and insect problems. It's the same as with humans, where the link between nutrition and susceptibility to diseases is well established. In addition, as the percent of organic matter in soil decreases, there is a decrease in the kinds of organisms present. Without competition, a disease organism can proliferate and cause more damage to crops. So it should come as no surprise that a healthy soil, well supplied with organic matter and containing a diverse population of organisms, will produce healthier plants than a soil with depleted organic matter.

Soil organic matter plays a crucial role in global cycles that are of major importance to life. Depletion of soil organic matter accelerates CO_2 buildup in the atmosphere. It can also result in more water runoff, and thus less recharge of groundwater.

Sources

Allison, F. E. 1973. *Soil organic matter and its role in crop production.* Amsterdam: Scientific Publishing Co.

Brady, N. C. 1990. *The nature and properties of soils.* 10th ed. New York: Macmillan Publishing Co.

Follett, R. F., J. W. B. Stewart, and C. V. Cole, eds. 1987. *Soil fertility and organic matter as critical components of production systems*. Special Publication No. 19. Madison, Wis.: Soil Science Society of America.

Lucas, R. E., J. B. Holtman, and J. L. Connor. 1977. Soil carbon dynamics and cropping practices. In *Agriculture and Energy*, edited by W. Lockeretz, pp. 333–51. New York: Academic Press. See this source for the Michigan study on the relationship between soil organic matter levels and crop-yield potential.

Powers, R. F., and K. Van Cleve. 1991. Long-term ecological research in temperate and boreal forest ecosystems. *Agronomy Journal* 83:11–24. This reference compares the relative amounts of carbon in soils with that in plants.

Stevenson, F. J. 1986. *Cycles of soil: Carbon, nitrogen, phosphorus, sulfur, micronutrients*. New York: John Wiley & Sons. This reference also compares the amount of carbon in soils with that in plants.

Tate, R. L., III. 1987. *Soil organic matter: Biological and ecological effects*. New York: John Wiley & Sons.

4

• • •

Organic Matter Levels in Soils

The depletion of the soil humus supply is apt to be a fundamental cause of lowered crop yields.—J. H. Hills, C. H. Jones, and C. Cutler, 1908

The amount of organic matter in any particular soil is a result of a wide variety of environmental, soil, and agronomic influences. Some of these, such as climate and soil texture, are naturally occurring. Pioneering work on the effect of natural influences on soil organic matter levels was carried out in the U.S. about 50 years ago by Hans Jenny. Human activity also influences soil organic matter levels. Tillage, crop rotation, and manuring practices all can have profound effects on the amount of soil organic matter.

The amount of organic matter in soil is a result of all the additions and losses of organic matter that have occurred over the years. In this chapter we will look at why different soils have different organic matter levels. Anything that adds large amounts of organic residues to a soil may cause organic matter increases. But anything that causes soil organic matter to decompose more rapidly or be lost through erosion may cause organic matter depletion. The organic matter content of a soil is a result of the amount of additions *relative* to losses over the years. A more detailed discussion of soil organic matter dynamics is given in Chapter 13.

Natural Factors

Temperature

In the United States it is easy to see how temperature affects soil organic matter levels. Traveling from north to south, average hotter temperatures lead to less soil organic matter. As the climate gets warmer two things tend to happen. As long as rainfall is sufficient, the amount of vegetation that is produced annually by plants increases. But the rate of decomposition of organic materials in soils, the amount lost each year, also increases, because soil organisms work more efficiently and for more of the year. This increasing decomposition with warmer temperatures becomes the dominant influence determining soil organic matter levels.

Rainfall

Soil organic matter levels generally increase as average annual precipitation increases. This is probably caused by *both* increased residue additions and lowered losses under higher rainfall situations. As there is more rainfall, more water is available to plants and more plant growth results. So, as rainfall increases there are generally more residues returned to the soil from grasses or trees. On the other hand, soils resulting from high rainfall may have less soil organic matter decomposition than well-aerated soils. Soils in arid climates usually have low amounts of organic matter. In a very dry climate such as a desert there is little production of vegetation. While decomposition may also be very low when the soil is dry and organisms cannot function well, when rains do occur there will be a very rapid burst of decomposition of soil organic matter.

Soil Texture

Soils of heavier texture, containing high percentages of clay, tend to have naturally higher amounts of soil organic matter. The organic matter content of sands may be less than 1% while loams may have 2 to 3% and clays from 4 to more than 5%. The strong bonds that develop between clay and organic matter seem to protect organic molecules from attack and decomposition by microorganisms. In addition, heavier soils tend to have more restricted pore space and the lower oxygen status also decreases decomposition. The lower decomposition rate that occurs in heavier soils is probably the main reason that their organic matter levels are higher than lighter soil.

Soil Drainage

Sometimes the topography will determine how well a soil drains. Soils in depressions at the bottom of hills are commonly pretty wet because they receive water and sediments from above them. Soils may also have a layer in the subsoil that doesn't allow water to drain well. Decomposition of organic matter occurs more slowly in badly aerated soils, when oxygen is limiting or absent, than in well-aerated soils. For this reason, organic matter accumulates in wet soil environments. In a totally flooded soil one of the major structural parts of plants, lignin, doesn't decompose at all. The ultimate consequence of extremely wet or swampy conditions is the development of organic (peat or muck) soils, with organic matter contents of over 20%. If organic soils are artificially drained for agricultural or other uses, the soil organic matter will decompose very rapidly. When this happens, the elevation of the soil surface actually decreases. Some people in

Florida whose houses were originally level with the ground and have corner posts going down below the organic level can now park their cars under their homes.

Position in the Topography

The soils at the bottom of a slope are generally wetter due to runoff from up-slope, and organic matter is not decomposed as rapidly in these soils as in drier soils. In addition, there is the effect of topography. Soils on a steep slope will tend to have low amounts of organic matter because the topsoil is continually eroded. At the same time, eroded material rich in organic matter may accumulate in soils at the bottom of the slope instead of completely leaving the field. This is another reason why soils at the bottom of hills usually have much more organic matter than soils on the slopes.

Type of Vegetation

What grows on a soil over the years affects the soil organic matter level. The most dramatic differences occur when soils developed under grassland are compared with those developed under forests. On natural grasslands organic matter tends to accumulate in high amounts and to be well distributed within the soil. This is probably a result of the deep and extensive root systems of native grasses. Their roots have high "turnover" rates, for root death and decomposition occurs as new roots are formed. The high levels of organic matter in soils that were once in grassland explains why these are some of the most productive soils in the world. In forests, litter accumulates on top of the soil, and surface organic layers commonly contain over 50% organic matter. On the other hand, subsurface mineral horizons in forest soils may contain from less than 1 to about 4% organic matter.

Acid Soil Conditions

In general, soil organic matter decomposition is slower under acid soil conditions than at more neutral pH. In addition, acid conditions, by inhibiting earthworm activity, may also cause organic matter to be concentrated at the soil surface rather than distributed in the soil.

Human Influences

Soil erosion removes topsoil enriched in organic matter so that eventually only subsoils remain. Crop production obviously suffers when part or all of the most fertile layer of the soil is removed. Erosion is a natural process and occurs on almost all soils. However, agricultural practices can result in a rapid acceleration of erosion. Some soils naturally erode more easily than others and the problem is also greater in some regions than others. On a nationwide basis soil erosion causes huge economic losses. It is estimated that erosion in the United States is responsible for annual losses of one-half billion dollars in available nutrients and $18 billion in total soil nutrients. Unless erosion is very severe, a farmer may not even realize that there is a problem. But that doesn't mean that crop yields are not being decreased. In fact, yields may decrease by 5 to 10% when only moderate erosion occurs. Yields may suffer a decrease of 10 to 20% or more with severe erosion. The results of a study of three midwestern soils, shown in table 4.1, indicate that erosion greatly influences both organic matter levels and water holding ability. Greater amounts of erosion, taking away organic matter–rich topsoil, caused a decrease in the organic matter contents of these loamy and clayey soils. In addition, eroded soils were able to store less available water than soils experiencing little erosion.

Organic matter is also lost from soils when soil organisms decom-

Table 4.1. Effects of Erosion

Soil	*Erosion*	*Organic Matter (%)*	*Potential Available Water (%)*
Corwin	slight	3.03	12.9
	moderate	2.51	9.8
	severe	1.86	6.6
Miami	slight	1.89	16.1
	moderate	1.64	11.5
	severe	1.51	4.8
Morley	slight	1.91	7.4
	moderate	1.76	6.2
	severe	1.60	3.6

Source: Schertz, Moldenhauer, Franzmeier, and Sinclair, 1985, tables 1 and 2, pp. 12, 13. Used with permission of the American Society of Agricultural Engineers.

pose more organic materials during the year than are made up by the amount of residues added. Many farming activities influence the level of organic matter in soils. Human influences on soil organic matter are introduced below and are discussed in more detail in Chapter 13.

Tillage Practices

Tillage practices influence both the amount of topsoil erosion and the rate of decomposition of soil organic matter. Conventional plowing and disking a soil to prepare a smooth seedbed also breaks down natural soil aggregates and destroys large water conducting channels. The soil is left in a physical condition that allows both wind and water erosion to occur easily.

The more a soil is disturbed by tillage practices, the greater the potential breakdown of organic matter by soil organisms. During the early years of agriculture in the United States, when forests were cleared and the soil plowed in the East and prairies plowed up in the Midwest, soil organic matter decreased rapidly. In fact, the soils were literally mined of a valuable resource—organic matter. In the Northeast and Southeast, it was quickly recognized that fertilizers and soil amendments were needed to maintain soil productivity. In the Midwest, the deep, rich soils of the tall-grass prairies were able to maintain their productivity for a long time despite accelerated soil organic matter loss and significant amounts of erosion. The reason for this was their unusually high original levels of soil organic matter.

Rapid soil organic matter decomposition by soil organisms usually occurs when a soil is worked with a moldboard plow. Incorporating residues, breaking aggregates open, and fluffing up the soil allows microorganisms to work more rapidly. It's something like opening up the air intake on a wood stove, which lets in more oxygen and causes the fire to burn hotter. In Vermont we found a 20% decrease in organic matter after five years of growing corn on a clay soil that had previously been in sod for a long time. In the Midwest, cultivation of soils for about forty years has caused a 50% decline in soil organic matter.

With the current interest in reduced (conservation) tillage, growing row crops in the future may not have such a detrimental effect on soil organic matter. Conservation-tillage practices leave more residues on the surface and cause less soil disturbance than conventional moldboard plow and disk tillage. In fact, soil organic matter levels usually increase when *no-till* planting machinery places seeds in a narrow band of disturbed soil while leaving the soil between planting rows undisturbed. The rate of decomposition of soil organic matter is lower, because the soil is not drastically disturbed by plowing and disking. Residues accumulate on the surface because the soil is not inverted by plowing. Earthworm populations increase, bringing some of the organic matter deeper into the soil and creating channels that help water infiltrate into the soil. Decreased erosion also results from using conservation-tillage practices.

Crop Rotations and Cover Crops

At different stages in a rotation, different things may be happening. Soil organic matter may decrease, then increase, then decrease, and so forth. While row crops under conventional moldboard plow cultivation usually result in decreased soil organic matter, many legumes, grasses, or legume-grass sods tend to increase soil organic matter. The "turnover" of the roots of sod plants plus the lack of soil disturbance allow organic matter to accumulate in the soil. In addition, different types of crops result in different quantities of residues returned to the soil. When corn grain is harvested more residues are left in the field than after soybean or lettuce harvests. Harvesting the same crop in different ways leaves different amounts of residues. When only corn is harvested more residues remain in the field than when the entire plant is harvested for silage.

Soil erosion is greatly reduced and topsoil rich in organic matter is conserved when cover crops, or green manures, such as grass or legume hay are grown year-round. The extensive root systems of sod crops account for much of the reduction in erosion. So having sod crops as part of a cover crop rotation helps to maintain organic matter by reducing loss of topsoil and decomposition of residues as well as by building up organic matter by extensive residue addition.

Organic Amendment Use

An old practice that helps maintain or increase soil organic matter is to apply manures or other organic residues from off the field. A study in Vermont during the 1960s and '70s found that between 20 and 30 tons (wet weight, including straw or sawdust bedding) of dairy manure per acre were needed to maintain soil organic matter levels when silage corn was grown each year. This is equivalent to 1 to 1-1/2 times the amount produced by a large Holstein cow over the whole year. Manures differ in their composition as well as how they are stored and handled in the field. Different manures can have very different effects on soil organic matter and nutrient availability.

Organic Matter Distribution in Soil

In general, more organic matter is present near the surface than deeper in the soil (see figure 4.1). This is one of the main reasons that topsoils are so productive compared to subsoils exposed by erosion or by mechanical removal of surface soil layers. Much of the plant residues that eventually become part of the soil organic matter are from the aboveground portion of plants. When the plant dies or sheds leaves or branches, residues are deposited on the surface. While earthworms and insects help to incorporate the residues on the surface deeper into the soil, the greatest concentrations still remain within 1 foot of the surface.

Litter layers that commonly develop on the surface of forest soils may have very high organic matter contents (figure 4.1a). Plowing forest soils after removal of the trees incorporates the litter layers into the mineral soil. An agricultural soil derived from a light, sandy

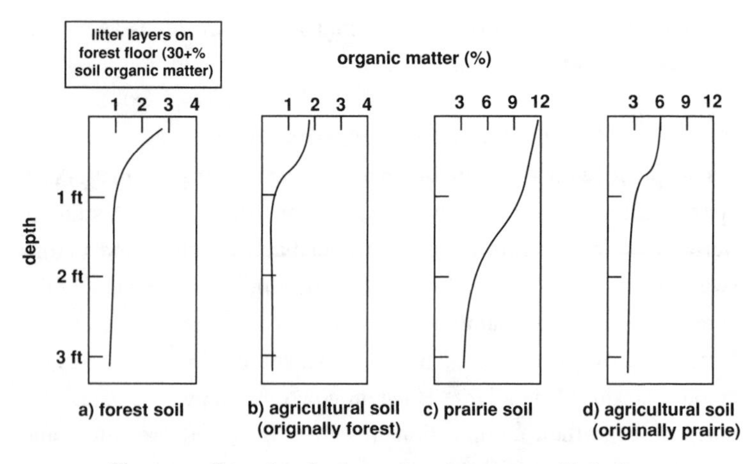

Figure 4.1. Examples of soil organic matter content with depth.

textured forest soil would likely have a distribution of organic matter similar to that indicated in figure 4.1b. Soils of the tall-grass prairies have high levels of organic matter deep into the soil profile (see figure 4.1c). After cultivation of these soils for fifty years there is far less organic matter present (figure 4.1d).

Summary

A number of natural factors combine to influence the amount of organic matter in soils. These factors have an influence on organic matter through their effect on plant growth and organic matter decomposition. Naturally high soil organic levels occur under conditions that support heavy production of vegetation, as in humid climates. Levels are also relatively high under conditions in which organic matter decomposition is restricted, as in cold climates or in high-clay-content soils. Organic matter tends to be higher in soils under grasslands than in those under forests.

Farming practices also influence soil organic matter levels. The types of plants grown, the use of manures, tillage practices, and crop rotations all affect the amount of soil organic matter. Practices that return large amounts of residues to the soil or reduce erosion promote relatively high organic matter levels.

Sources

Brady, N. C. 1990. *The nature and properties of soils.* 10th ed. New York: Macmillan Publishing Co.

Carter, V. G., and T. Dale. 1974. *Topsoil and civilization.* Norman: University of Oklahoma Press.

Hass, H. J., G. E. A. Evans, and E. F. Miles. 1957. *Nitrogen and C changes in Great Plains soils as influenced by cropping and soil treatments.* United States Department of Agriculture Technical Bulletin 1164. Washington, D.C.: U.S. Government Printing Office. This is a reference for the large decrease in organic matter content of Midwest soils.

Jenny, H. 1980. *The soil resource.* New York: Springer-Verlag.

Jenny, H. 1941. *Factors of soil formation.* New York: McGraw-Hill. Jenny's early work on the natural factors influencing soil organic matter levels.

Magdoff, F. R., and J. F. Amadon. 1980. Yield trends and soil chemical changes resulting from N and manure application to continuous corn. *Agronomy Journal* 72:161–64. See this reference for further information on the studies in Vermont cited in this chapter.

National Research Council. 1989. *Alternative agriculture.* Washington, D.C.: National Academy Press.

Schertz, D. L., W. C. Moldenhauer, D. F. Franzmeier, and H. R. Sinclair, Jr. 1985. Field evaluation of the effect of soil erosion on crop productivity. In *Erosion and soil productivity.* Proceedings of the national symposium on erosion and soil productivity. American Society of Agricultural Engineers Publication 8–85. St. Joseph, Michigan.

Tate, R. L., III. 1987. *Soil organic matter: Biological and ecological effects*. New York: John Wiley & Sons.

Part Two: Practices

5

• • •

Organic Matter Management—the Balancing Act

Because organic matter is lost from the soil through decay, washing, and leaching, and because large amounts are required every year for crop production, the necessity of maintaining the active organic-matter content of the soil, to say nothing of the desirability of increasing it on many depleted soils, is a difficult problem.—A. F. Gustafson, 1941

It is difficult to be sure exactly why problems develop when organic matter is depleted in a particular soil. However, even at the turn of the century soil scientists proclaimed, "Whatever the cause of soil unthriftiness, there is no dispute as to the remedial measures. Doctors may disagree as to what causes the disease, but agree as to the medicine. Crop rotation! The use of barnyard and green manuring! Humus maintenance! These are the fundamental needs" (Hills, Jones, and Cutler, 1908). More than eighty years later these are still the major remedies available to us.

There seems to be a contradiction in our view of soil organic matter. On the one hand, we want crop residues, dead microorganisms, and manures, to decompose. If there is no decomposition of soil organic matter then no nutrients can be made available to plants, no glue to bind particles can be manufactured, and no humus can be produced to hold on to plant nutrients as water leaches through the soil. On the other hand, we know that numerous problems develop when soil organic matter is significantly depleted through decomposition. This dilemma of wanting organic matter to decompose but not wanting to lose too much means that organic materials must be continually added to the soil. A supply of "active" organic matter must be maintained so that humus can continually accumulate. This does not mean that organic materials must be added to each field every year. However, it does mean that a field cannot go without additions of organic residues for many years without paying the consequences.

Do you remember that plowing a soil is similar to opening up the air intake on a wood stove? What we really want in soil is a slow "steady burn" of the organic matter. You get that in a wood stove by adding wood every so often and making sure the air intake is on a medium setting. In soil, you get a steady burn by adding organic residues regularly and by not disturbing the soil too often.

Three things will help you to maintain or increase soil organic matter: control of erosion, addition of organic residues, and reduction of the rate at which organic matter is decomposed by organisms. Soil erosion must be controlled to keep organic matter–enriched topsoil in place. In addition, organic matter added to a soil must either match or exceed the rate of loss by decomposition. These additions can come from manures and composts brought from off the field, crop residues remaining following harvest, or cover crops. Reduced tillage lessens the rate of organic matter decomposition and may also result in less erosion. It is not possible to give specific soil organic matter management recommendations for all situations. However, in Chapters 6 through 11 some management options are evaluated in detail and issues associated with their use are discussed.

Soil Organic Matter Levels

Raising and Maintaining Soil Organic Matter Levels

It is not easy to dramatically increase the organic matter content of soils or even to maintain good levels once they are reached. It can only be done with a sustained effort that includes a number of approaches to returning organic materials to soils and rotating with high-residue crops. It is especially difficult to raise the organic matter content of soils that are very well aerated, such as coarse sands, because added materials are decomposed so rapidly. Soil organic matter levels can be maintained with less organic residue in high-clay-content soils with restricted aeration than in coarse-textured soils.

How Much Is Enough?

There are a number of reliable soil tests to tell if a field is sufficiently supplied with plant nutrients such as phosphorus, potassium, nitrogen, and magnesium. Soil tests can also show whether a soil is too acid, contains potentially toxic levels of aluminum, or contains too much sodium. We can easily measure the organic matter content of soils or the amount of a particular part of the organic matter such as humic acids. Unfortunately, there is no accepted interpretation for these organic matter tests, no guidelines as to what a test result means for a particular soil. We *do* know some general guidelines. For example, 2% organic matter in a sandy soil is very good, but in a clay soil 2% indicates a greatly depleted situation. But the complexity of soil organic matter composition, including biological diversity of organisms as well as the actual organic chemicals present, means that there is no simple interpretation for the various tests.

Residue Characteristics

Some residues, such as fresh, young, and very green plants, decompose rapidly in the soils and in the process may readily release plant nutrients. This could be compared to the effect of sugar eaten by humans, which results in a quick burst of energy. Some of the substances in older plants and in the woody portion of trees, such as lignin, decompose very slowly in soils. Well-composted organic residues also decompose slowly in the soil because they are fairly stable, having already undergone a significant amount of decomposition.

The ratio of the amount of a residue's carbon to the amount of its nitrogen influences nutrient availability and the rate of decomposition. The ratio, usually referred to as the *C:N ratio,* may vary from around 20:1 for young plants, between 50 to 80:1 for the old straw of crop plants, to over 100:1 for sawdust. For comparison, the C:N ratio of soil organic matter is usually in the range of about 10 to 13:1 and the C:N of soil microorganisms is around 5 to 7:1.

The C:N ratio of residues is really just another way of looking at the percentage of nitrogen (figure 5.1). A high C:N residue has a low percentage of N. Low C:N residues have relatively high percentages of N. Crop residues are usually pretty close to 40% carbon and this figure doesn't change that much from plant to plant. On the other hand, nitrogen content varies greatly depending on the type of plant and its stage of growth.

Mature plant stalks and sawdust that have C:N over 40:1 can cause temporary problems for plants. Microorganisms using materials containing 1% or less N need extra nitrogen for their growth and reproduction. They will take the needed nitrogen from the surrounding

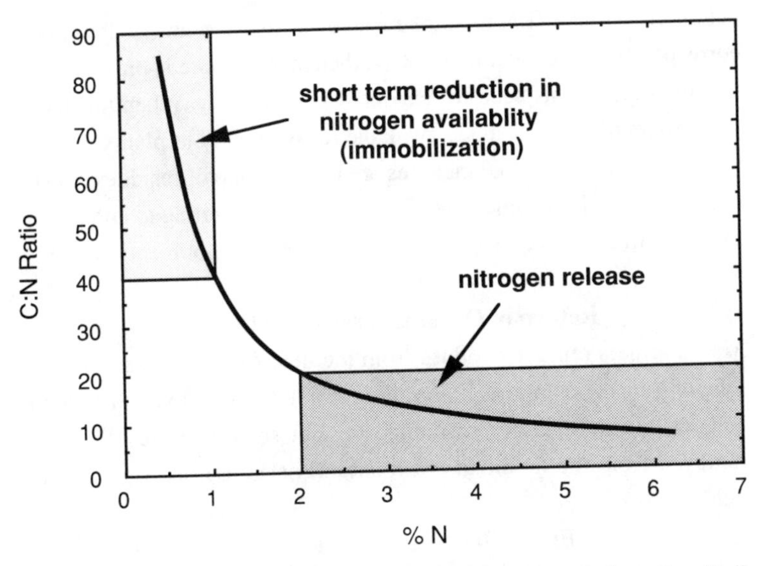

Figure 5.1. The relationship of % N and C:N ratio of crop residues. Redrawn from Vigil and Kissel, 1991, p. 759. Courtesy of the Soil Science Society of America, Inc.

soil, diminishing the amount of nitrate and ammonium available for crop use. This reduction of soil nitrate and ammonium by microorganisms decomposing high C:N residues is called immobilization of nitrogen.

When microorganisms and plants compete for scarce nutrients, the microorganisms usually win because they are so well distributed in the soil. Plant roots are in contact with only 1 to 2% of the entire soil volume while microorganisms populate almost the entire soil. The length of time during which the nitrogen nutrition of plants is adversely affected by immobilization depends on the quantity of residues applied, their C:N ratio, and other factors influencing microorganisms such as fertilization practices and temperature and moisture conditions. If the C:N ratio of residues is in the teens or low 20s, corresponding to greater than 2% N, then there is more N present than the microorganisms need for residue decomposition. When this happens, extra nitrogen is made available fairly quickly to plants. Green manure crops and animal manures are in this group of residues. Residues with C:N in the mid-20s to low 30s, corresponding to about 1 to 2% N, will not have much effect on nitrogen immobilization or release.

Return of Organic Residues to the Soil

Many farmers remove residues from the field for use as animal bedding or to make compost. Later, these residues return to contribute to soil fertility as manures or composts. But sometimes residues are removed from the field and not returned, or are burned right in the field.

Field Burning of Residues

Burning of wheat, rice, and other crop residues in the field is a common practice in parts of the United States as well as in other countries. Residues are usually burned to help control insects or diseases or to make next year's field work easier. Residue burning may be so widespread in a given area that it causes a local air-pollution problem. Burning also diminishes the amount of organic matter returned to the soil.

Competition for Organic Materials

There are sometimes important needs for crop residues and manures that may prevent their use in maintaining or building soil organic matter. For example, straw may be removed from a grain field to serve as mulch in a strawberry field. These trade-offs of organic materials can sometimes cause a severe soil-fertility problem, especially in developing countries where resources are scarce. There, crop residues and manures frequently serve as fuel for cooking or heating when gas, coal, oil, or wood are not available. In addition, straw may be used as thatch for housing or to make fences. While it is completely understandable that people in resource-poor regions use residues for such purposes, the negative effects of these uses on soil productivity can be substantial. An important way to increase agricultural productivity in developing countries is to find alternative sources for fuel and building materials to replace the crop residues and manures traditionally used.

Application Rates for Organic Materials

Many times the amount of residue added to a soil will be determined by the cropping system. The crop residues can be left on the surface or incorporated by tillage. Different amounts of residue will remain under different crops, rotations, or harvest practices. For example, 3 tons per acre of leaf, stalk, and cob residues remain in the field when corn is harvested for grain. If the entire plant is harvested to make silage, there is next to nothing left except the roots.

When "imported" organic materials are brought to the field you need to decide how much to apply. In general, application rates of these residues will be based on their probable contribution to the nitrogen nutrition of plants. We don't want to apply too much available nitrogen because it will be wasted. Nitrate from excessive applications of organic sources of fertility may leach into groundwater just as easily as nitrate originating from purchased synthetic fertilizers. In addition, excess nitrate in plants may cause health problems for humans and animals.

Sometimes the fertility contribution of phosphorus may be the main factor governing application rates of organic material. There is concern that excess phosphorus entering some lakes is causing an increase in the growth of algae and other aquatic weeds. This decreases the quality of the water for drinking as well as recreation. In these locations, care needs to be taken to avoid loading the soil with too much phosphorus from either commercial fertilizers or organic sources.

Effect of Residue and Manure Accumulations

When any organic material is added to soil, it decomposes relatively rapidly at first. But later, when only resistant parts, like straw stems high in lignin, are left, the rate of decomposition decreases greatly. What this means is that although nutrient availability diminishes each year after adding a residue to the soil, there are still long-term benefits from adding organic materials. This can be expressed by using a "decay series." For example, 50%, 15%, 5%, and 2% of the amount of nitrogen added in manure may be released in the 1st, 2d, 3d, and 4th years following an addition to soil. In other words, crops in a regularly manured field get some nitrogen from manure that was applied in past years. So, if you are starting to manure a field, somewhat more manure will be needed in the first year than will be needed in years 2, 3, and 4 to supply the same total amount of nitrogen to a crop each year.

Management Systems

Animal-based Systems

It is certainly easier to maintain soil organic matter in animal-based agricultural systems. Manure is a valuable by-product of having animals. Animals can also use sod-type grasses and legumes as pasture, hay, and haylage. It is easier to justify putting land into perennial forage crops for part of a rotation when there is an economic use for the crops. Animals need not be on the farm to have positive effects on soil fertility. A farmer may grow hay to sell to a neighbor and trade for some animal manure from the neighbor's farm, for example. Occasionally, formal agreements between dairy farmers and vegetable growers lead to cooperation on crop rotations and manure application.

Systems without Animals

It is more challenging, although not impossible, to maintain or increase soil organic matter on non–livestock farms. It can be done by using cover crops, intercropping, living mulches, rotations that include crops with high amounts of residue left after harvest, and attention to other erosion-control practices. Organic residues such as leaves or sewage sludges can sometimes be obtained from nearby cities and towns. Straw or grass clippings used as mulch also add organic matter when they later become incorporated into the soil by plowing or by the activity of soil organisms. Some vegetable farmers are experimenting with a "mow-and-blow" system where crops are grown on strips for the purpose of chopping them and spraying the residues onto an adjacent strip.

Maintaining a Diverse Environment

The role of diversity is important in maintaining a well functioning and stable ecological system. Where many different types of organisms coexist in the same area there are fewer disease, insect, and nematode problems. There is more competition for food and more possibility that many types of predators will also be found. This means that no single pest organism will be able to reach a population high enough to cause a major decrease in crop yield. We can promote a diversity of plant species growing on the land by using cover crops, intercropping, and crop rotations. But don't forget that diversity below the soil surface is as important as diversity aboveground. Growing cover crops and using crop rotations help maintain the diversity belowground, but adding manures and composts and making sure that crop residues are returned to the soil are also critical in promoting soil organism diversity.

Summary

Although we don't know the exact amount of organic matter needed in a particular soil for the best production of crops, it is clear that productivity suffers when organic matter levels decline. Soil organic matter management involves using a number of possible strategies. Reducing soil erosion to a minimum will decrease the loss of organic matter in the eroded topsoil. Regular additions of organic materials like manures, composts, green manures, and crop residues make up for losses of soil organic matter by decomposition. The characteristics of organic materials may determine application rates.

Soil organic matter levels can be maintained more easily if animals are part of the management system. This is because of the availability of manures and the economic benefits that result from growing sod crops. Growing legume and grass sod crops as part of a rotation helps to lessen erosion and increase soil organic matter. To maintain or increase organic matter without animals, greater dependence must be placed on practices such as cover cropping, growing crops with a lot of residues, and erosion control.

Sources

Brady, N. C. 1990. *The nature and properties of soils.* 10th ed. New York: Macmillan Publishing Co.

Vigil, M. F., and D. E. Kissel. 1991. Equations for estimating the amount of nitrogen mineralized from crop residues. *Soil Science Society of America Journal* 55:757–61.

6

• • •

Animal Manures

The quickest way to rebuild a poor soil is to practice dairy farming, growing forage crops, buying . . . grain rich in protein, handling the manure properly, and returning it to the soil promptly.—J. L. Hills, C. H. Jones, and C. Cutler, 1908

Once cheap fertilizers became widely available after World War II, many farmers, extension agents, and scientists looked down their noses at manure. People thought more about how to get rid of manure than how to put it to good use. In fact, scientists tried to find out the absolute maximum amount of manure that could be applied to an acre without reducing crop yields. Some farmers who didn't want to spread manure sometimes actually piled it next to a stream and hoped that next spring's flood waters would wash it away. We now know that manure, like money, is better spread around than concentrated in a few places. Talking of money, the economic contribution of farm manures can be considerable. The value of the nutrients in manure from a 70-cow dairy farm may exceed $7,000 per year, while manure from a 50-sow farrow-to-finish operation is worth about $4,000 and manure from a 20,000-bird broiler operation is worth about $3,000. If you're not getting the full fertility benefit from manures on your farm, you may be wasting plenty of money.

Animal manures can have very different properties depending on the animal species, feed, bedding, and manure-storage practices. The amount of nutrients in the manure that become available to crops also depends on what time of the year the manure is applied and how quickly it is worked into the soil. In addition, the influence of manure on soil organic matter and plant growth is influenced by soil type. In other words, it's impossible to give specific manure application recommendations for every situation.

We'll deal mainly with dairy cow manure because there's more information about its use on cropland. General information will also be given about the characteristics and uses of some other animal manures.

Manure Handling Systems

Solid versus Liquid

The type of barn on the farmstead may determine how manure is handled on a dairy farm. Dairy-cow manure containing a fair amount of bedding can be spread as a solid even though it contains about 80 to 87% water. This is most common on farms where cows are kept in individual stanchions. Liquid manure-handling systems are common where animals are kept in a "free stall" barn with little bedding. Liquid manure is usually 91% or more water. Manures that have characteristics that are between solid and liquid are usually referred to as semi-solid or slurry, depending on the method of handling.

Storage of Manure

Researchers have been investigating how best to store manure to reduce the problems that come with year-round manure spreading. Storage means that the farmer can apply manure when it's best for the crop. When manure is stored, farmers are not forced into spreading it in any weather. This can reduce nutrient loss from the manure caused by water runoff from the field. However, significant losses of nutrients from stored manure also may occur. One study found that during the year dairy manure stored in uncovered stacks, or piles, lost 3% of the solids, 10% of the nitrogen, 3% of the phosphorus, and 20% of the potassium. Covered stacks or well-contained liquid systems, which tend to form a crust on the surface, do a better job of conserving the nutrients and solids in manures than unprotected stacks. Poultry manure, with its high amount of ammonium, may lose 50% of its nitrogen during storage as ammonia gas volatilizes, unless precautions are taken to conserve nitrogen.

Chemical Characteristics of Manures

A high percentage of the nutrients in feeds passes right through animals and ends up in manures. Over 70% of the nitrogen, 60% of the phosphorus, and 80% of the potassium in feeds may be available in manures for recycling onto cropland. In addition to the N, P, and K contributions given in Table 6.1, manures also contain significant amounts of other nutrients such as calcium, magnesium, and sulfur. In regions where the micronutrient zinc tends to be deficient, there is rarely any deficiency on soils receiving regular manure applications.

The values given in Table 6.1 must be viewed with some caution because the characteristics of manures from even the same type of animal may vary considerably from one farm to another. Differences in feeds, mineral supplements, bedding materials, and storage systems make manure analysis quite variable. But on a given farm, as long as feeding, bedding, and storage practices remain unchanged, manure characteristics will be similar from year to year.

The major difference among all the manures is that poultry manure is significantly higher in nitrogen and phosphorus than the other ma-

Table 6.1. Manure Characteristics

	Dairy cow	*Beef cow*	*Chicken*	*Pig*
DRY-MATTER CONTENT (%)				
Solid (fresh)	13	12	25	9
Liquid (fresh, diluted)	9	8	17	6
TOTAL NUTRIENT CONTENT (APPROXIMATE)				
Nitrogen				
lb/ton	10	14	25	10
lb/1,000 gal.	28	39	70	28
Phosphate (P_2O_5)				
lb/ton	5	9	25	6
lb/1,000 gal.	14	25	70	9
Potash (K_2O)				
lb/ton	10	11	12	9
lb/1,000 gal.	28	31	33	25
MANURE EQUIVALENTS				
Solid manure (tons)	20	11	5	16
Liquid manure (gal.)	7,200	4,000	1,500	5,700

Source: Based on Madison et al. 1986.

nure types. This is partially due to the difference in feeds given poultry and other farm animals. The higher percentage of dry matter in poultry manure compared to other manures is also partially responsible for the higher analyses of certain nutrients when expressed on a wet ton basis.

It is possible to take the guesswork out of estimating manure characteristics—most soil-testing laboratories will now analyze manure. Manure analysis should become a routine part of the soil-fertility management program on animal-based farms.

Applying Manures

Manures, like other organic residues that decompose easily and rapidly release nutrients, are usually applied to soils in quantities judged to supply sufficient nitrogen for the crop being grown in the current year. While it might be better for building and maintaining soil organic matter to apply manure at higher rates, doing so may cause undesirable nitrate accumulation in leafy crops and excess nitrate leaching to groundwater. While high nitrate levels in leafy-vegetable crops may be undesirable in terms of human health, high nitrate in the leaves of many plants also seems to make them more attractive to insects. In addition, salt damage to crop plants can occur from high manure application rates, especially when there is insufficient leaching by rainfall or irrigation. It is also a waste of money and resources to add unneeded nutrients to the soil, nutrients which will only be lost by leaching or runoff instead of contributing to crop nutrition.

A common per-acre rate of dairy-manure application is 10 to 30 tons fresh weight, or 4,000 to 11,000 gallons of liquid manure. These rates will supply approximately 50 to 150 pounds of available nitrogen (not total) per acre. If you are growing crops that don't need that much nitrogen, such as small grains, 10 to 15 tons should supply sufficient N per acre. Low rates of about 10 tons per acre are also suggested for each of the multiple applications used on a hay crop. For a crop that needs a lot of nitrogen such as corn, 20 to 30 tons per acre may be necessary to supply its N needs.

For the most nitrogen benefit to crops, manures should be incorporated into the soil immediately after spreading on the surface. About half of the total nitrogen in dairy manure comes from the ammonium (NH_4^+) in urine. This ammonium represents almost all of the readily available nitrogen present in dairy manure. As materials containing ammonium dry on the soil surface, the ammonium is converted to ammonia gas (NH_3) and lost to the atmosphere. If dairy manure stays on the soil surface, about 25% of the nitrogen is lost after one day and 45% is lost after 4 days. This problem is significantly lessened if rainfall occurs shortly after manure application, leaching ammonium from manure into the soil. Leaving manure on the soil surface is also a problem because runoff waters may carry significant amounts of nutrients from the field. When this happens, crops don't benefit as much from the manure application, and surface waters become polluted.

Other nutrients contained in manures, in addition to nitrogen, make important contributions to soil fertility. The availability of phosphorus and potassium in manures should be similar to that in commercial fertilizers. The phosphorus and potassium contributions of 20 tons of dairy manure is approximately equivalent to about 30 to 50 lbs of phosphate and 180 to 200 lbs of potash from fertilizers. Trace elements in manure, such as the zinc previously mentioned, also add to the fertility value of this resource.

Potential Problems

As we all know, too much of a good thing is not necessarily good. Excessive manure applications may cause plant-growth problems. It is especially important not to apply excess poultry manure because the high soluble-salt content can harm plants.

Plant growth is sometimes retarded when high rates of *fresh* manure are applied to soil immediately before planting. This problem usually doesn't appear if the fresh manure decomposes for a few weeks in the soil, and can also be avoided by using a solid manure that has been stored for a year or more. Injection of liquid manure sometimes causes problems when used on poorly drained soils in wet years. It may be that the extra water applied and the extra use of oxygen by microorganisms means less aeration for plant roots.

When manures are applied regularly to a field in order to provide enough nitrogen for a crop like corn, phosphorus may build up to levels way in excess of crop needs. Erosion of phosphorus-rich topsoils contributes sediments to streams and lakes, sediments believed to be a major contributing factor to surface-water pollution. In cases of very high phosphorus buildup in soils resulting from the continual application of manure, it may be wise to switch application to other fields or to use strict soil-conservation practices to trap sediments before they enter a stream.

Farms that purchase much of their animal feed may have too much manure to safely use on their own land. Although they don't usually realize it, they are importing large quantities of nutrients in the feed that remain on the farm as manures. If they apply all these nutrients on a small area of soil, nitrate pollution of groundwater will occur. Both nitrate and phosphate pollution of surface waters can also occur. It is a good idea to make arrangements with neighbors for use of the excess manure. Another option, if local outlets are available, is to compost the manure (see Chapter 10) and sell the product to garden centers, landscapers, and directly to home gardeners.

Effects of Manuring on Soil Organic Matter

When considering the influence of any residue or organic material on soil organic matter, the key question is the amount of solids returned to the soil. Equal amounts of different types of manures will have different effects on soil organic matter levels. Dairy and beef manure containing undigested parts of forages as well as bedding have a high amount of complex substances, such as lignin, that do not decompose readily in soils. Using this type of manure will result in a greater long-term influence on soil organic matter than will a poultry manure without bedding. More solids are commonly applied to soil with solid manure-handling systems than with liquid systems because greater amounts of bedding are usually included.

When conventional tillage is used to grow a crop such as corn silage, where the entire aboveground portion will be harvested, research has indicated that an annual application of 20 to 30 tons of the solid type of dairy manure per acre is needed to maintain soil organic matter. As discussed above, a nitrogen-demanding crop such as corn may be able to use all of the nitrogen in 20 to 30 tons of manure. If more residues are returned to the soil by just harvesting grain, lower rates of manure application will be sufficient to maintain or build up soil organic matter.

One large Holstein "cow year" worth of manure is about 20 tons. While 20 tons of anything is a lot, when considering dairy manure it translates into a much smaller amount of solids. If the approximately 5,200 pounds of solid material in the 20 tons is applied over the surface of one acre and mixed with the 2,000,000 pounds of soil present to a 6-inch depth, it would raise the soil organic matter by about 0.3%. However, much of the manure will decompose during the year, so the *net* effect on soil organic matter will be even less. Let's assume that 75% of the solid matter decomposes during the first year and that 25% of the original 5,200 pounds, or 1,300 pounds, is therefore added to the stable part of soil organic matter. The net effect would then be an increase in soil organic matter of 0.065% (the calculation is [1,300/2,000,000]x100). While this does not seem like much added organic matter, if a soil had 2.17% organic matter and 3% of this was decomposed annually during cropping, then the loss would be 0.065% per year and the manure addition would just balance this loss.

Summary

Manures are valuable by-products of animal agriculture. They contain plant nutrients and if well-utilized can significantly reduce the need to purchase fertilizers. Manures also play an important role in maintaining or increasing soil organic matter. But in order to have a major influence on soil organic matter content, large quantities of manures must be applied.

The chemical characteristics of manures vary widely, depending upon the type of animal, feed and mineral supplements given to the animal, bedding, and how the manure is handled. Manure analysis should become an important part of the soil-fertility management program on farms using significant amounts of manures.

Sources

Elliott, L. F., and F. J. Stevenson, eds. 1977. *Soils for management of organic wastes and wastewaters.* Madison, Wis.: Soil Science Society of America.

Madison, F., K. Kelling, J. Peterson, T. Daniel, G. Jackson, and L. Massie. 1986. *Guidelines for applying manure to pasture and cropland in Wisconsin.* Agricultural Bulletin A3392.

Magdoff, F. R., and J. F. Amadon. 1980. Yield trends and soil chemical changes resulting from N and manure application to continuous corn. *Agronomy Journal* 72:161–64. See this reference for dairy manure needed to maintain or increase organic matter under continuous cropping for silage corn.

Magdoff, F. R., J. F. Amadon, S. P. Goldberg, and G. D. Wells.

1977. Runoff from a low-cost manure storage facility. *Transactions of the American Society of Agricultural Engineers* 20:658–60, 665. This is the reference for the nutrient loss that can occur from uncovered manure stacks.

Maryland State Soil Conservation Committee. N.d. *Manure management handbook—a producer's guide.*

Soil Conservation Society of America. 1976. *Land application of waste materials.* Ankeny, Iowa: Soil Conservation Society of America.

7

• • •

Cover Crops

Where no kind of manure is to be had, I think the cultivation of lupines will be found the readiest and best substitute. If they are sown about the middle of September in a poor soil, and then plowed in, they will answer as well as the best manure.—Columella, first century, Rome

The understanding of the effect of crops on the soil and on the productivity of following crops comes down to us from antiquity. Chinese manuscripts indicate that the use of green manures is probably 3 thousand years old. Green manures were also commonly used in ancient Greece and Rome. There are three different terms used to describe crops grown specifically to help maintain soil fertility and productivity instead of for harvesting: green manures, cover crops, and catch crops. The terms are sometimes used interchangeably and can be best thought of as describing the main goal of the grower. A green manure crop is usually grown to help maintain soil organic matter as well as to increase nitrogen availability. A cover crop is grown mainly to help retard soil erosion by keeping the ground covered with living vegetation and by having living roots holding onto the soil. This of course is related to managing soil organic matter, because the topsoil lost during erosion contains the most organic matter of any soil layer. A catch crop is grown to retrieve available nutrients still in the soil following an economic crop and keeps them from being leached over the winter.

Sometimes it's confusing to decide which term to use—green manure, cover crop, or catch crop. We usually have more than one goal when we plant these crops during or after our main crop. And whether we realize it or not, plants grown for one of these purposes many times also accomplish the other two goals. The question of which term to use is not really important, so in our discussion below the term *cover crop* will be used.

Cover crops are usually incorporated into the soil or killed on the surface before they are mature. (This is the origin of the term *green* manure.) Since cover crop residues are usually low in lignin content and high in nitrogen, with a low C:N ratio, they decompose rapidly in the soil.

Effects of Cover Crops

Cover crop residues replenish organic matter that is naturally decomposed during the year. The benefits from cover crops depend a lot on how long the crop is left to grow before the soil is prepared for the next crop.

The more residues you return to the soil, the better the effect on soil organic matter. Cover crops will increase soil organic matter only if they're allowed a long enough growing period for high dry-matter production. The amount of residues produced by the growth of a green manure can be very small, sometimes around 1/2 ton of dry matter per acre. But good production of hairy vetch or crimson clover cover crops may yield 1-1/2 to 2-1/2 tons per acre. If a crop like cereal rye is grown to maturity it can produce 3 to 5 tons of residues.

A five-year experiment with clover in California showed that cover crops increased organic matter in the top 2 inches from 1.3 to 2.6% and in the 2 to 6 inch layer from 1.0 to 1.2%. On the other hand, some researchers have found that green manures do not seem to increase soil organic matter. Even if they don't, cover crops will help prevent erosion and will add some new residues.

Cover crops can also have other important effects, supplying nutrients to the following crop, suppressing weeds, and breaking pest cycles. Their pollen and nectar can be important food sources for predatory mites and parasitic wasps, both important for biological control of insect pests. A cover crop also provides a good habitat for spiders, and those insect feeders will help decrease populations of pests. Living cover crop plants as well as their residues also increase water infiltration into soil. This can help compensate for the water that cover crops use.

Selection of Cover Crops

There are some questions about cover crops that you need to answer before growing them: Which type should be planted? When and how should the crop be planted? When should the crop be killed or incorporated into the soil?

When you select a cover crop you should consider what you want to accomplish, the soil conditions, and the climate. Is the main purpose to add available nitrogen to the soil or to provide large amounts of organic residue? Is erosion control in the late fall and early spring the primary objective? Is weed suppression the main goal? Is the soil very acid and infertile, with low availability of nutrients? Does it have a compaction problem? Think about the climate, too, because some species are more winter-hardy than others. In addition, the climate and water-holding properties of your soil may determine whether or not a cover crop uses so much water that it harms the following crop.

Legumes and grasses, including cereals, have been the most extensively used green manures. Leguminous crops are often very good cover crops. Summer annual legumes, usually grown only during the summer, include soybeans, peas, and beans. Winter annual legumes that are normally planted in the fall and counted on to overwinter include berseem clover, crimson clover, hairy vetch, and subterranean clover. Some, like crimson clover, can only overwinter in regions with mild frost. Hairy vetch, though, is able to withstand fairly severe winter weather. Biennials and perennials include red clover, white clover, sweet clover, and alfalfa. It should be noted that crops which are usually used as winter annuals are sometimes grown as summer annuals in cold, short-season regions. Also, summer annuals that are easily damaged by frost, such as cowpeas, can be grown as a winter annual in the deep South.

One of the main reasons for selecting legumes as cover crops is their provision of available nitrogen to the soil. A legume such as hairy vetch or crimson clover that produces a substantial amount of growth may supply over 100 pounds of nitrogen per acre to the next crop. However, other legumes such as field peas, bigflower vetch, and red clover may supply only 30 to 80 pounds of available nitrogen.

Nonleguminous crops used as cover crops include the cereal grasses rye, wheat, oats, and barley, as well as other grass-family species such as ryegrass. Other cover crops, like buckwheat, rape, and turnips, are neither legumes nor grasses.

There are many different types of plants that can be used as cover crops. Some of the important ones currently used are discussed below.

Legumes

If you grow a legume as a cover crop don't forget to inoculate seeds with the bacteria that live in the roots and fix nitrogen. There are various types of rhizobia bacteria that fix nitrogen. Some will work only with alfalfa. Other groups of plants having their own special types of rhizobia are the clovers, soybeans, beans, peas and vetch (both need the same type of bacteria), and cowpeas. If you have not recently grown a legume from the same general group you are going to plant, it would be worthwhile to mix the seeds with a commercial rhizobia inoculum before planting. You can add some thick sugar water to the seed-inoculum mix in order to get bacteria to stick better to the seeds. The inoculum that you need will be readily available only if it is commonly used in your region. It's best to check with your seed supplier a few months before you need the inoculum so that it can be special ordered if necessary.

Winter Annuals

Berseem clover is an annual crop that is grown in the South during the winter. Newly released varieties have done very well in California, with "Multicut" outyielding "Bigbee." It establishes easily and rapidly and develops a dense cover, making it a good choice for weed suppression. It is also drought tolerant and regrows rapidly when mowed or grazed.

Bigflower vetch is winter-hardy but has generally not performed as well as hairy vetch. However, bigflower vetch is resistant to vetch weevil and anthracnose, both of which can sometimes reduce hairy vetch's usual advantage. Bigflower vetch flowers and completes its life cycle a few weeks earlier than hairy vetch and can therefore eliminate the need to use herbicides to kill the cover crop in time to plant the summer main crop.

Crimson clover has long been considered one of the best cover crops for the southeastern United States. Where adapted, it will grow in the fall and winter and mature more rapidly than most other legumes. It will also contribute more nitrogen to the following crop than other cover crops. Since it is not very winter-hardy, crimson clover is not usually a good choice even in the northern portion of the South. In northern regions, crimson clover can be grown as a summer annual, but then no economic crop can be grown during that field season. A few varieties such as Chief, Dixie, and Kentucky Select are somewhat winter-hardy if established early enough before winter. Crimson clover does not grow well on high pH (calcareous) or poorly drained soils.

Hairy vetch is grown in the Southeast, but is winter-hardy enough to grow well in the Mid-Atlantic states and even in most of the Northeast and Midwest. Where adapted, hairy vetch will produce a large amount of vegetation and fix a significant amount of nitrogen, contributing as much as 100 pounds or more of nitrogen per acre to the next crop. Hairy vetch residues decompose rapidly and release nitrogen more quickly than most other cover crops. This can be an advantage when a rapidly growing, high-nitrogen-demand crop follows hairy vetch. Hairy vetch will do better on sandy soils than many other green manures.

Subterranean clover is a warm climate winter annual that, in many situations, can complete its life cycle before a summer crop is planted. When used this way it doesn't need to be suppressed or killed and will not compete with the summer crop. Because it grows low to the ground and does not tolerate much shading it is not a good choice to interplant with summer annual row crops.

Summer Annuals

Cowpeas are native to central Africa and do well in hot climates. The cowpea is, however, severely damaged by even a mild frost. It is deep rooted and is able to do well under droughty conditions. It will usually do better on low-fertility soils than crimson clover.

Soybeans, usually grown as an economic crop for their oil and protein-rich seeds, can also serve as a summer green manure crop. They require a fertile soil for best growth. As with cowpeas, soybeans can be easily damaged by frost.

Biennials and Perennials

Alfalfa is a good choice for well-drained soils, near neutral in pH, and high in fertility. The good soil conditions required for the best growth of alfalfa mean that it is not very useful for many problem situations. Where adapted, it is usually grown in a rotation for a number of years (see chapter 8). Alfalfa is commonly interseeded with small grains such as oats, wheat, and barley and then allowed to grow after the grain is harvested. There has been considerable interest in the use of a relatively new alfalfa variety, Nitro, as a cover crop. Nitro is not winter-hardy, and behaves as an annual under northern conditions. It is able to continue fixing nitrogen later in the fall than winter-hardy varieties. Since it is winter-killed, herbicides are not needed when Nitro is followed by no-till corn.

Crown vetch is only adapted to well-drained soils, but can be grown under lower fertility conditions than alfalfa. It has been used successfully for roadbank stabilization and is able to provide permanent groundcover. Crown vetch has been tried as a "living mulch" interseeding with only limited success at providing N to corn. However, it is relatively easy to suppress crownvetch with herbicides to reduce its competition with corn.

Red clover is vigorous, shade tolerant, and winter-hardy and can be established relatively easily. Red clover is commonly interseeded with small grains. Because red clover starts growing slowly, the competition between it and the small grain is not usually great. Red clover has also been used successfully as an interseeding in corn in the Northeast.

Sweet clover (yellowblossom) is a reasonably winter-hardy, vigorous-growing crop with an ability to get its roots into compacted subsoils. It is able to withstand high temperatures and droughty conditions better than many other cover crops. It requires a soil pH near neutrality and a high calcium level. But as long as the pH is high, sweet clover is able to grow pretty well on low-fertility soils. It is sometimes grown for a full year or more, since it flowers and completes its life cycle in the second year. When used as a green manure crop it is incorporated into the soil before full bloom.

White clover does not produce as much growth as many of the other legumes and is also less tolerant of droughty situations. However, because it does not grow very tall and is able to tolerate shading better than many other legumes, it may be useful in orchard-floor covers or as a living mulch. It is also a common component of intensively managed pastures.

Grasses

A problem common to all the grasses is that if you grow the crop to maturity for the maximum amount of residue, you reduce the amount of available nitrogen for the next crop. This is caused by the high C:N ratio, or low percentage of N, in grasses near maturity. The problem can be avoided by killing the grass early or by adding extra available nitrogen in the form of fertilizer or manure. Another way to help with this problem is to supply extra nitrogen by seeding a legume-grass mix.

Winter rye, also called cereal or grain rye, is very winter-hardy and easy to establish. Its ability to germinate quickly, together with its winter-hardiness, means that it can be planted later in the fall than most other species. Winter rye has been shown to have an *allelopathic* effect, which means that it can chemically suppress weeds. It grows quickly in the fall and also grows readily in the spring.

Oats are not winter-hardy. Summer or fall seedings will winter-kill under most northern conditions. This provides a naturally killed mulch the following spring and may help in weed suppression. As a mixture with one of the clovers, it can provide some quick growth in the fall and its stems can help trap snow and conserve moisture even after it has been killed by frost.

Annual ryegrass can grow well in the fall if established early enough. It also develops a very extensive root system and can therefore provide very effective erosion control while adding significant quantities of organic matter. It may winter-kill in northern climates. Some caution is needed with annual ryegrass because it can become a problem weed in some situations.

Sudangrass and sorghum-sudan hybrids are fast-growing summer annuals that can produce a lot of growth in a short time. Because of their vigorous nature they are good at suppressing weeds. If they are used as an interseeding with a low-growing crop such as strawberries or many vegetables, you may need to delay seeding so that your main crop will not be severely shaded.

Other Crops

Buckwheat is a summer annual that is easily killed by frost. It will grow better than many other cover crops on low-fertility soils. It also grows rapidly and completes its life cycle quickly. Buckwheat can grow more than 2 feet tall in the month following planting. It competes well with weeds because it grows so fast and therefore can be used to suppress weeds following an early spring vegetable crop. It is possible to grow more than one crop of buckwheat per year in many regions. Its seeds do not disperse widely and it can reseed itself. Mow it or till it under before seeds develop if you don't want it to reseed.

Rape is a winter-hardy member of the Crucifer (cabbage) family. It grows well under the moist and cool conditions of late fall, while other kinds of plants just sit there and get ready for winter. Rape will be killed by harsh winter conditions in the North, but can be grown as a winter crop in the middle and southern sections of the country.

Mixtures of Cover Crops

Mixtures of cover crops are planted so that benefits are realized from two or more different species. The most common mixture is a grass and legume, such as winter rye and hairy vetch or oats and red clover. Mixed stands will usually do a better job of suppressing weeds than a single species. Growing legumes with grasses also helps deal with the problem discussed earlier, where turning under grasses near maturity may decrease nitrogen availability for the following crop. In addition, a crop that grows erect, such as winter rye, may provide support for hairy vetch and enable it to grow better. Mowing close to the ground can kill vetch supported by rye easier than vetch alone. This can allow mowing to replace herbicides in no-till production systems.

Timing Cover Crop Growth

If you want to accumulate a lot of organic matter, it's best to grow a cover crop for the whole growing season (see figure 7.1a). This means

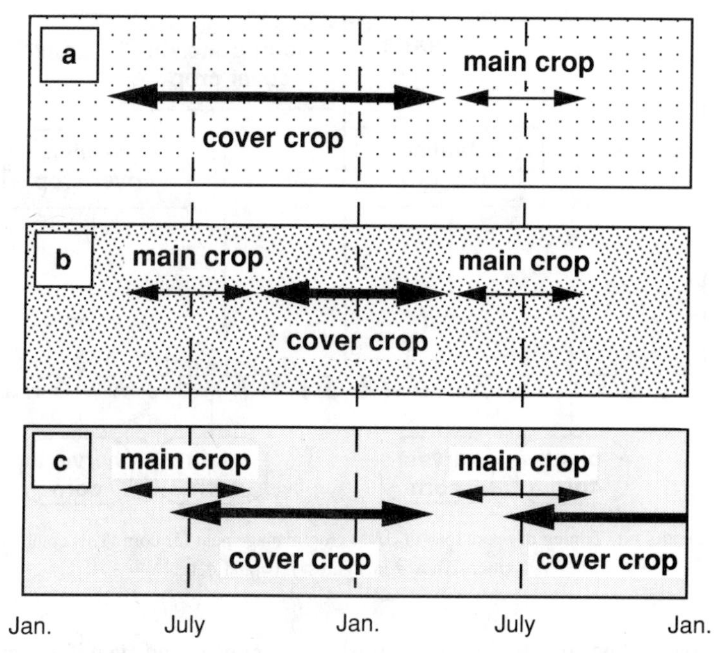

Figure 7.1. Three ways to time cover crop growth for use with a summer crop.

there will be no income-generating crop grown that year. This may be useful with very infertile, possibly eroded, soils. It also may be helpful in vegetable production systems when there is no manure available and where a market for hay crops justifies a longer rotation.

Most farmers sow cover crops after the economic crop has been harvested (figure 7.1b). In this case, as with the system shown in figure 7.1a, there is no competition between the cover crop and the main crop. The seeds can be drilled instead of broadcast, resulting in better cover crop stands. In the Deep South you can usually plant cover crops after harvesting the main crop. But in northern areas there

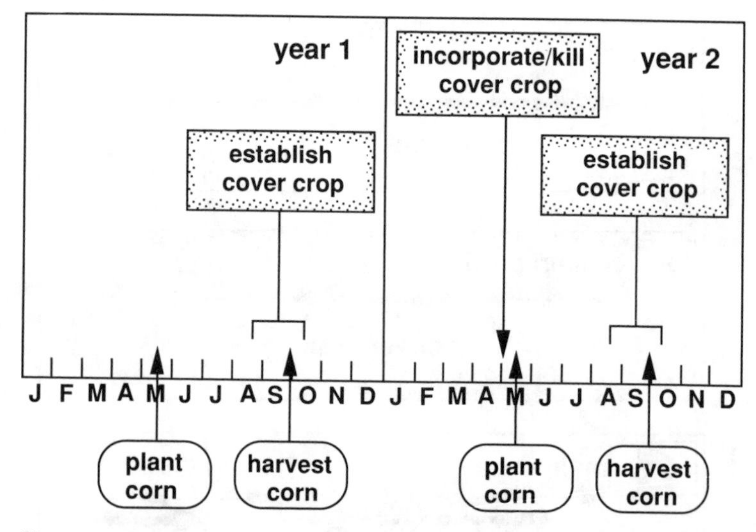

Figure 7.2. Timing of operations in cover crop management for corn. This combines options shown in figures 7.1b and 7.1c.

may not be enough time to establish a cover crop. And even if you are able to get it established, there will be little growth in the fall to provide soil protection or nutrient uptake. The choice of cover crop is severely limited in northern climates by the short growing season and severe cold. Winter rye is probably the most reliable cover crop for this situation.

The third management strategy is to interseed cover crops during the growth of the main crop (figure 7.1c). This system is especially helpful in the establishment of cover crops in short-growing-season areas. Delay seeding the cover crop until the main crop is off to a good start and will be able to grow well despite the competition. Good establishment of cover crops requires moisture and, for small-seeded crops, some covering of the seed by crop residues. On the other hand,

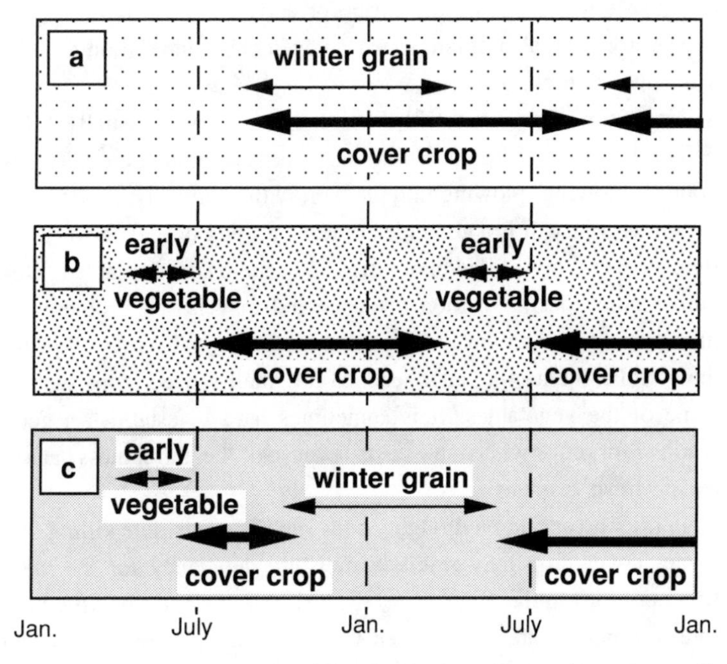

Figure 7.3. Timing cover crop growth for winter grain, early vegetable, and vegetable-grain systems.

cereal rye is able to establish well without seed covering as long as sufficient moisture is present. Farmers using this system usually broadcast seed during or just after the last cultivation. Some have used aerial seeding and "highboy" tractors or detasseling machines to broadcast green manure seed after a main crop is already fairly tall.

When used with winter grain cropping systems, cover crops are established following grain harvest in late spring or interseeded with the grain during fall planting (figure 7.3a). With some early-maturing vegetable crops, especially in warmer regions, it is also possible to establish cover crops in late spring or early summer (figure 7.3b). Cover crops can also fit into an early vegetable–winter grain rotation sequence (figure 7.3c).

No matter when you establish cover crops, they are usually killed before or during soil preparation for the next economic crop. This is done by mowing, plowing into the soil, with chemicals, or naturally by winter injury. It is a good idea to leave a week or two between the time a cover crop is tilled in or killed and a main crop is planted. This allows some decomposition to occur and may lessen problems of nitrogen immobilization and allelopathic effects. It also may allow for the establishment of a better seedbed for small-seeded crops such as some of the vegetables. It is sometimes hard to establish a good seedbed for crops with small seeds because of the clumpiness caused by the fresh residues.

In drier areas and on droughty soils such as sands, late killing of a winter cover crop may result in moisture deficiency for the main summer crop. In these situations the cover crop should be killed before too much water is removed from the soil. However, in warm climates where no-till methods are practiced, it actually might be better to let the cover crop grow longer. More growth will mean more residues and better water conservation from the mulch when the main crop is growing. This better mulch may more than compensate for the extra water removed from the soil during the later period of green manure growth.

In very humid regions or on wet soils the ability of an actively growing cover crop to "pump" water out of the soil by transpiration may be an advantage. Letting the cover crop grow for as long as possible will result in more rapid soil drying and allow for earlier planting of the main crop.

Cover crops are sometimes allowed to flower to provide bees or other beneficial insects with pollen. However, if the plants actually set seed the cover crop may be reseeded unintentionally. The possibility of a cover crop later becoming a weed can be a problem with crops such as ryegrass and hairy vetch.

Intercrops

Growing a cover crop between the rows of a main crop has been practiced for a long time. It has been called a living mulch, an intercrop, polyculture (if more than one crop will be harvested), and an orchard-floor cover. This practice has many benefits. Compared to a "clean" soil, a ground cover provides erosion control, better conditions for using equipment for harvesting crops, higher water-infiltration capacity, and an increase in soil organic matter. In addition, if the cover crop is a legume there can be a significant buildup of nitrogen that may be available to crops in future years. Another benefit is the attraction of beneficial insects such as predatory mites to flowering plants. This may help explain the finding of less insect damage under polyculture than under monoculture.

Growing other plants near the main crop also poses potential dangers. The intercrop may harbor insect pests, such as the tarnished plant bug. Most of the management decisions in using intercrops are connected with trying to minimize competition with the main crop. Intercrops, if they grow too tall, can compete with the main crop for light. An intercrop can also physically interfere with the main crop's growth or harvest. There is also a potential competition for water and nutrients. Using intercrops is a highly questionable practice if rainfall is barely adequate for the main crop and supplemental irrigation isn't available. One way to decrease competition is to delay seeding the intercrop until the main crop is well established. This is sometimes done in commercial fruit orchards. With main crops that are annuals there is not much difference between delayed plantings of soil-improving intercrops and cover crops. Herbicides, mowing, and partial rototilling have been used to suppress the cover crop and give an advantage to the main crop. Another way to lessen competition from the cover is to plant the main crop in a relatively wide cover-free strip. This provides more distance between the main crop and the intercrop rows.

Summary

Cover crops have a large number of potentially beneficial effects on soils and growing crops. They are effective in protecting the soil against erosion and the resulting loss of topsoil. They also help maintain or build up organic matter through the addition of residue, make the soil more porous, increase water infiltration, and they may add nitrogen to the soil and help control weeds and insect pests. Numerous types of plants, both legumes and nonlegumes, can serve as highly useful cover crops under certain conditions.

There are a number of ways that cover crops can fit into agricultural production systems. The main difference is in the extent of overlap between the growth of the cover crop and main economic crops. Cover crops may be planted after harvest of the main crop or interseeded during all or part of the growth of the main crop. The longer the extent of overlap, the greater the potential for reduced yields of the main crop through competition.

Sources

Allison, F. E. 1973. *Soil organic matter and its role in crop production.* Amsterdam: Elsevier Scientific Publishing Co. In his discussion of organic matter replenishment and green manures (pp. 450–51), Allison cites a number of researchers who indicate that there is little or no effect of green manures on total organic matter, even though the supply of active (rapidly-decomposing) organic matter increases.

Brusko, M., and the staff of the Rodale Institute. 1992. *Managing cover crops profitably.* Sustainable Agriculture Network, Handbook Series, #1. USDA Sustainable Agriculture and Education Program.

Hargrove, W. L., ed. 1991. *Cover crops for clean water.* Ankeny, Iowa: Soil and Water Conservation Society.

MacRae, R. J., and G. R. Mehuys. 1985. The effect of green manuring on the physical properties of temperate-area soils. *Advances in Soil Science* 3:71–94.

Miller, P. R., W. L. Graves, W. A. Williams, and B. A. Madson. 1989. *Covercrops for California agriculture.* Leaflet 21471. Division of Agriculture and Natural Resources, University of California. This is the reference for the experiment with clover in California.

Pieters, A. J. 1927. *Green manuring principles and practices.* New York: John Wiley & Sons.

Power, J. F., ed. 1987. *The role of legumes in conservation tillage systems.* Ankeny, Iowa: Soil Conservation Society of America.

Smith, M. S., W. W. Frye, and J. J. Varco. 1987. Legume winter cover crops. *Advances in Soil Science* 7:95–139.

8

• • •

Crop Rotations

Grass is a source of strength of agriculture and, therefore, to the nation. The more we fail to realize this, the more difficult it will be to maintain and build up our great agricultural resources, our soil resources, and, yes, our human resources, too.—Henry Wallace, 1940

There are very good reasons to rotate the crops you grow in each field. With rotations there are usually fewer problems with insects, parasitic nematodes, weeds, and diseases caused by bacteria, viruses, and fungi. Rotations are effective in controlling insects such as the corn rootworm and diseases such as root rot of field peas. In addition, when legumes are part of the rotation they supply nitrogen to succeeding crops. Growing sod-type grasses, legumes, and grass-legume mixes as part of the rotation also increases soil organic matter. When you alternate two crops, such as corn and soybeans, you have a very simple rotation. More complex rotations require three or more crops and a five- to ten-year (or more) cycle to complete.

Yields of crops grown in rotations are frequently higher than when grown in monoculture, where the same crop follows year after year. When you grow a grain or vegetable crop following a legume, the extra supply of nitrogen can help a lot. But yields of crops grown in rotation are often higher than in monoculture even when both are supplied with plentiful amounts of nitrogen. In addition, following a nonlegume crop with another nonlegume can produce higher yields than monoculture. For example, when you grow corn following rye, or cotton following corn, you get higher yields than when corn or cotton are grown alone. This yield benefit from rotations is sometimes called a *rotation effect* in scientific articles. Another important benefit of rotations is that growing a variety of crops in a given year instead of only one or two spreads out labor needs and reduces risk caused by climate or market conditions. Because of all these reasons, rotations are an important part of any sustainable agricultural system.

Two things happen when sod crops are part of the rotation and remain in place for some years during a rotation. First, the rate of decomposition of soil organic matter decreases because the soil is not continually being disturbed. (This also happens when using no-till planting, even for non–sod-type crops such as corn.) Second, grass and legume sods develop extensive root systems, part of which will naturally die each year, adding new organic matter to the soil. Older roots of grasses die even during the growing season and provide sources of "fresh" organic matter. Roots of plants also continually give off, or exude, a variety of chemicals that nourish nearby microorganisms. In these ways, crops with extensive root systems stimulate high levels of soil biological activity.

We are interested in the living portion of the organic matter and want many different organisms living in the soil. We also want to have a good general level of soil organic matter, or humus, in the soil. Although there have been many experiments performed that compared soil organic matter changes under different cropping systems, few experiments have looked at the effects of rotations on soil ecology.

Soil Biology

The more residues your crops leave in the field the greater the populations of soil microorganisms. Experiments in a semiarid region in Oregon found that the total amount of microorganisms in a two year wheat-fallow system was only about 25% of the amount found under pasture. Conventional moldboard plow tillage systems are known to decrease the populations of earthworms as well as other soil organisms. So when your rotations are more complex and include sod crops you will increase soil biological diversity.

Soil Organic Matter Levels

You might think you're doing pretty well if soil organic matter remains the same under a particular cropping system. But if you are working soils with depleted organic matter you need to build up levels to counter the effects of previous practices. Maintaining an inadequately low level of organic matter is not exactly a wonderful achievement!

The types of crops you grow, the amount of roots they have, their yield, the portion of the crop that is harvested, and how you treat the crop residues will all affect soil organic matter. Soil fertility itself will influence the amount of organic residues returned, since more fertile soils grow higher-yielding crops, resulting in more residues.

Compared to the original forest or prairie soil, organic matter levels are usually lower under conventional cultivation for field crops such as corn, wheat, soybeans, cotton, and potatoes. The decrease is more rapid, and fertility problems become worse, if no manures are

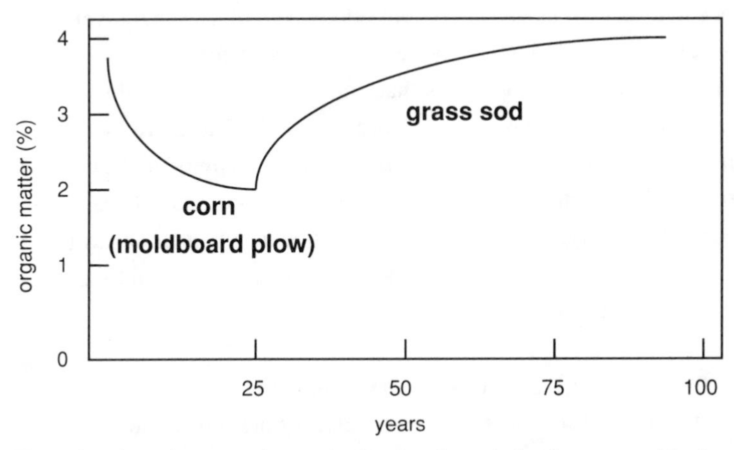

Figure 8.1. Organic matter changes in the plow layer during long-term cultivation followed by sod establishment.

applied. Erosion is a greater problem under agricultural systems than under natural vegetation. In addition, added organic residues are usually not sufficient to make up for losses resulting from decomposition by soil organisms. The decrease is very rapid for the first five to ten years, but eventually a plateau, or equilibrium, is reached. After that, soil organic matter levels remain pretty much the same as long as practices aren't changed. An example of what can occur during twenty-five years of continuously grown corn is given in figure 8.1. However, soil organic matter levels increase when the cropping system is changed from a cultivated crop to a grass or mixed grass-legume sod. The increase is usually slower than the decrease that occurred under continuous tillage.

A long-term cropping experiment in Missouri compared continuous corn to continuous sod and various rotations. More than 9 inches of topsoil was lost during sixty years of continuous corn. The amount of soil lost *each year* from the continuous-corn plots was equivalent to 21 tons per acre. After sixty years, soil under continuous corn had only 44% as much topsoil as that under continuous timothy sod. A six-year rotation consisting of corn-oats-wheat-clover-two years of timothy resulted in about 70% as much topsoil as found in the timothy soil, a much better result than with continuous corn. Differences in erosion and organic matter decomposition resulted in soil organic matter proportions of 2.2% for the unfertilized timothy and 1.2% for the continuous-corn plots.

Residue Availability

As pointed out in Chapter 5, more residues are left in the field after some crops than others. For example, more residues remain following corn grown for grain than following soybeans, corn silage, cabbage, or lettuce.

For crops such as lettuce, cabbage, and corn silage, almost the entire aboveground portion of the plant is removed from the field at harvest. This leaves few residues to help replenish soil organic matter. On the other hand, when corn is harvested for grain, about 50% of the aboveground plant matter remains following harvest. Harvesting corn for grain may therefore leave 3 to 4 tons of residue per acre more than harvesting for silage. If residues are left on the surface, there will usually be less erosion.

Farm Labor and Economics

Before discussing appropriate rotations, let's consider some of the possible effects on farm labor and finances. If only one or two row crops are grown, farmers must work incredibly long hours during planting and harvesting seasons. But including sod-type crops and early harvested crops along with those that are traditionally harvested in the fall allows farmers to spread their labor over the growing season, making labor needs more manageable. In addition, when you grow a more diversified group of crops, you are less affected by price fluctuations of one or two crops. This provides more year-to-year financial stability.

General Principles

Try to consider the following principles when you're thinking about a new rotation:

1. Follow a legume-sod crop with a high-nitrogen-demanding crop such as corn to take advantage of the nitrogen supply.

2. Grow less-nitrogen-demanding crops such as oats, barley, or wheat in the second or third year after a legume sod.

3. Grow the same annual crop for only one year if possible to decrease the likelihood of insects, diseases, and nematodes becoming a problem.

4. Don't follow one crop with another closely related species, since insect, disease, and nematode problems are frequently shared by members of closely related crops.

5. Use crop sequences that promote healthier crops. Some crops seem to do well following a particular crop—cabbage family crops following onions, or potatoes following corn. Other crop sequences may have adverse effects, as when potatoes have more scab following peas or oats.

6. Use crop sequences that aid in controlling weeds. Small grains seem to compete strongly against weeds and may inhibit germination of weed seeds, row crops permit mid-season cultivation, and sod crops that are mowed regularly or intensively grazed help control annual weeds.

7. Use longer periods of perennial crops, such as a legume sod, on sloping land and on highly erosive soils. Using sound conservation

practices, such as no-till planting or strip-cropping along the contour, may lessen the need to follow this guideline.

8. Try to grow a deep-rooted crop, such as alfalfa, safflower, or sunflower as part of the rotation. It can scavenge deep in the soil for nutrients and water, and channels left from decayed roots can promote water infiltration.

9. Grow some crops that will leave a significant amount of residues, like sorghum or corn harvested for grain, to help maintain organic matter levels.

Rotation Examples

It's impossible to recommend specific rotations for a wide variety of situations. Every farm has its own unique combination of soil and climate, of human and animal and machine resources. The economic conditions and needs are also different on each farm. But you may get some useful ideas by considering a number of rotations with historical or current importance.

A five to seven year rotation was common in the mixed livestock-crop farms of the northern Midwest and Northeast during the first half of the twentieth century. An example of this rotation is the following:

- one year of corn
- one year of oats
- three years of a mixed grass-legume hay
- two years of pasture

The most nitrogen-demanding crop, corn, followed the pasture, and grain was harvested only two of every five to seven years. A less nitrogen-demanding crop, oats, was planted in the second year as a "nurse crop" when the grass-legume hay was seeded down. The grain was harvested as animal feed and oat straw was harvested to be used as cattle bedding, but both eventually were returned to the soil as animal manure. This rotation was able to maintain soil organic matter in many situations, or at least didn't cause it to decrease too much. On prairie soils, with their very high original contents of organic matter, levels still probably decreased with this rotation.

In the corn belt region of the Midwest a change in rotations occurred as pesticides and fertilizers became readily available and animals were fed in large feedlots instead of on crop-producing farms. Once the mixed livestock farms became grain-crop farms or crop-hog farms, there was little reason to grow sod crops. In addition, federal commodity-price-support programs unintentionally encouraged farmers to narrow production to just two feed grains. The two-year corn-soybean rotation is better than monoculture, but it has a number of problems, including erosion, groundwater pollution with nitrate and herbicides, and depletion of soil organic matter. Whereas no-till practices can reduce soil erosion, they won't reduce pollution by agricultural chemicals. Research indicates that with high yields of corn grain there may be sufficient residues to maintain organic matter. But soybean residues are minimal.

An alternate five-year corn belt rotation, practiced on the Thompson farm in Iowa, a mixed crop-livestock (hogs and beef) operation, is similar to the first rotation described. For fields that are convenient for pasturing beef cows the rotation is as follows:

- one year of corn
- one year of soybeans
- one year of corn
- one year of oats (mixed legume/grass hay seeded)
- three years of meadow

Organic matter is maintained through a combination of practices that include the use of manures and sewage sludge, green manure crops (oats and rye following soybeans and hairy vetch following corn), crop residues, and sod crops. These practices have resulted in a porous soil that has significantly lower erosion than neighbors' fields.

A four-year rotation under investigation in Virginia uses mainly no-till practices as follows:

Year 1. Corn planted and harvested, winter wheat no-till planted into corn stubble

Year 2. Winter wheat harvested, fox-tail millet no-till planted into wheat stubble and hayed or grazed, alfalfa no-till planted in fall

Year 3. Alfalfa harvested and/or grazed

Year 4. Alfalfa harvested and/or grazed as usual until fall, then heavily stocked with animals to weaken it so that corn can be planted the next year

This rotation follows many of the principles we discussed earlier in this chapter. It was designed by researchers, extension specialists, and farmers. It's very similar to the older rotation described earlier, except that it is shorter, alfalfa is used instead of clover or clover-grass mixtures, and there is a special effort to minimize pesticide use under no-till practices. Weed-control problems have occurred when going from alfalfa (fourth year) back to corn. This has caused the investigators to use fall tillage followed by a cover crop mixture of winter rye and hairy vetch. They have had some success in suppressing the cover crop in the spring by just rolling over it with a disk harrow and planting corn through the surface residues with a modified no-till planter. The heavy cover crop residues on the surface have provided excellent weed control for the corn.

Traditional wheat-cropping patterns for the semi-arid regions of the Great Plains and the Northwest commonly include a fallow year to allow storage of water for use by the next wheat crop. But the wheat-fallow system has several problems. Because no crop residues are returned during the fallow year, soil organic matter decreases unless manure or other organic materials are provided from off the field. Water infiltrating below the root zone during the fallow year moves salts through the soil to the low parts of fields. Shallow groundwater can come to the surface in these low spots and create saline "seeps" where yields will be decreased. This system may also cause enough leaching of salts to degrade the groundwater of the region. Increased soil erosion caused by either wind or water commonly occurs. The build-up of grassy-weed populations such as jointed goat grass and downy brome under the wheat monoculture system also indicates that crop diversification is essential.

Farmers in this region who are trying to develop more sustainable cropping systems should consider using a number of species, including deeper-rooted crops, in a more complex rotation; increasing the amount of residues returned to the soil; reducing tillage; and lessening or eliminating the fallow period.

A four-year wheat-corn-millet-fallow rotation under evaluation in Colorado has been found to be better than the traditional wheat-fallow system. Wheat yields have been higher in this rotation than wheat grown in monoculture. The extra residues from the corn and millet are also helping to increase soil organic matter. Sunflower, a deep-rooting crop, is being evaluated in Oregon as part of a wheat-cropping sequence.

A ten-year rotation developed for vegetable farms and gardens by Eliot Coleman is as follows:

Year 1. Snap beans and radishes
Year 2. Carrots and onions
Year 3. Greens
Year 4. Potatoes
Year 5. Corn

Year 6. Peas
Year 7. Broccoli/cauliflower
Year 8. Winter squash/summer squash/cucumbers
Year 9. Tomatoes and green peppers
Year 10. Corn

Each year a cover crop is undersown into the vegetable crop.

Another rotation for vegetable growers uses a two-to-three-year alfalfa sod as part of a six-to-eight-year cycle. In this case the crops following the alfalfa would be high-nitrogen-demanding crops such as corn or squash, followed by cabbage or tomatoes, and in the last two years, crops needing a fine seedbed, such as lettuce, onions, or carrots. Annual weeds in this rotation are controlled well by the many harvests made each year during the alfalfa phase of the rotation. Perennial weed populations can be decreased by cultivation during the row-crop phase of the rotation.

Most vegetable farmers do not have enough land to have a multi-year legume sod on a significant portion of their land. However, judicious use of cover crops can help maintain organic matter. Manure and/or composts should also be applied every year or two to help maintain soil organic matter. Composting (see Chapter 10) of crop residues is usually convenient in a home garden or on a small farm, but unfortunately hasn't yet been used much on larger commercial farms.

Home gardeners can maintain soil organic matter by using a bagging lawn mower to provide grass clippings for mulch during the growing season. The mulch can then be worked into the soil or left on the surface to decompose until the next spring.

Summary

Effective crop rotations are an important part of sustainable agricultural systems. In addition to the positive effects on soil organic matter, rotations help break disease, weed, and insect-pest cycles. When crops are grown in a rotation, they usually do better than when the same crop is grown in the same field year after year. Rotations can also spread out labor needs more evenly during the year. By growing a larger number of crops, farmers are also better protected from the economic effects of large swings in the price of one or two crops. A number of principles should be followed when building rotations for a particular farm.

Sources

Anderson, S. H., C. J. Gantzer, and J. R. Brown. 1990. Soil physical properties after 100 years of continuous cultivation. *Journal of Soil and Water Conservation* 45:117–21.

Baldock, J. O., and R. B. Musgrave. 1980. Manure and mineral fertilizer effects in continuous and rotational crop sequences in central New York. *Agronomy Journal* 72:511–18.

Coleman, E. 1989. *The new organic grower.* Chelsea, Vt.: Chelsea Green. See this reference for the vegetable rotation.

Francis, C. A., and M. D. Clegg. 1990. Crop rotations in sustainable production systems. In *Sustainable Agricultural Systems,* ed. C. A. Edwards, R. Lal, P. Madden, R. H. Miller, and G. House. Ankeny, Iowa: Soil and Water Conservation Society.

Gantzer, C. J., S. H. Anderson, A. L. Thompson, and J. R. Brown. 1991. Evaluation of soil loss after 100 years of soil and crop management. *Agronomy Journal* 83:74–77. This source describes the long-term cropping experiment in Missouri.

Havlin, J. L., D. E. Kissel, L. D. Maddux, M. M. Claassen, and J. H. Long. 1990. Crop rotation and tillage effects on soil organic carbon and nitrogen. *Soil Science Society of America Journal* 54:448–52.

Luna, J. M., V. G. Allen, W. L. Daniels, J. F. Fontenot, P. G. Sullivan, C. A. Lamb, N. D. Stone, D. V. Vaughan, E. S. Hagood, and D. B. Taylor. 1991. Low-input crop and livestock systems in the southeastern United States. In *Sustainable agriculture research and education in the field,* pp. 183–205. Proceedings of a conference, April 3–4, 1990. Board on Agriculture, National Research Council. Washington, D.C.: National Academy Press. This is the reference for the rotation experiment in Virginia.

National Research Council. 1989. *Alternative agriculture.* Washington, D.C.: National Academy Press. This is the reference for the rotation used on the Thompson farm.

Peterson, G. A., and D. G. Westfall. 1990. Sustainable dryland agroecosystems. In *Conservation tillage.* Proceedings of the Great Plains conservation tillage system symposium, August 21–23, 1990, Bismark, North Dakota. Great Plains Agricultural Council Bulletin No. 131. See this reference for the wheat-corn-millet-fallow rotation under evaluation in Colorado.

Rasmussen, P. E., H. P. Collins, and R. W. Smiley. 1989. *Long-term management effects on soil productivity and crop yield in semi-arid regions of eastern Oregon.* USDA–Agricultural Research Service and Oregon State University Agricultural Experiment Station, Columbia Basin Agricultural Research Center, Pendleton, Oregon. This describes the Oregon study of sunflowers as part of a wheat-cropping sequence.

Werner, M. R., and D. L. Dindal. 1990. Effects of conversion to organic agricultural practices on soil biota. *American Journal of Alternative Agriculture* 5(1): 24–32.

9

• • •

Reduced Tillage

A plowboy and two horses lined against the gray,
Plowing in the dusk the last furrow.
The turf had a gleam of brown,
And smell of soil was in the air,
And, cool and moist, a haze of April.
—Carl Sandburg, early 1900s

How you till the soil and prepare the seedbed can have profound effects on the quality of your soil. Farmers till for many reasons: they want to create a good seedbed, control weeds, and bury crop residues and manures. In this chapter we'll focus on how different tillage practices affect soil organic matter. A word of caution—soils and climates and needs of farmers are so different there won't be one tillage system that works best everywhere.

The Moldboard Plow

The moldboard plow is the standard implement used around the world in the first step to preparing soils for planting. It may come as a surprise to learn that the modern moldboard plow can be traced back to plows used in China three thousand years ago. The introduction of the moldboard plow in Holland and England in the seventeenth century, along with the technology of grain drills (also of Chinese origin), greatly contributed to the agricultural revolution that in turn made possible the industrial revolution in Europe. The moldboard plow, which cuts under the soil and completely inverts it, also contributed significantly to agriculture in the United States. In the era before herbicides and modern no-till equipment, working the grassland soils of the great prairies, with their thick sods, would have been unthinkable without the moldboard plow.

The Conventional Tillage System

The traditional tillage system developed in the United States starts with plowing to a depth of 8 to 10 inches with a moldboard plow in the fall on wet soils or in the spring on better-drained soils. Soils are then disked in the spring once or twice to further break down aggregates and smooth the soil surface before planting. During the growing season a cultivator is sometimes used between the rows of widely spaced crops for weed control and/or to incorporate fertilizer.

This type of system works reasonably well in Europe, where erosion is less of a problem than in the United States and rotations that include sod crops are commonly used. In the United States, some of our best soils are very prone to erosion. For example, the loess soils that are widespread in the prairie region were formed from windblown materials that were held in place by the dense prairie vegetation. Once plowed up, wind and the rain produced by intense prairie rainstorms can move the soil particles fairly easily. The dust bowl of the 1930s, centered in the region where Texas, Oklahoma, Kansas, Colorado, and New Mexico converge, was an eyeopener to farmers and conservationists alike. The dramatic dust storms in this region, and the severe water erosion that etched deep gullies into the landscape in other parts of the United States, forced people to question the agricultural practices that were then being used. It became apparent that some of the cultivated soils in the dust bowl region should have remained in prairie vegetation. But improper tillage and residue-management practices had also contributed greatly to erosion.

Conventional tillage decreases soil organic matter and increases the potential for erosion by wind and water. There are two ways this happens:

1. Plowing leaves no residues on the surface to lessen the impact of falling rain and provide temporary storage for water during rainfall. Earthworms can find few food sources at the surface and so populations of certain species decrease drastically. The opening up and aeration of the soil plus the burying of residues permits organisms to decompose organic matter more rapidly.

2. Tillage smoothes the surface and destroys natural soil aggregates and channels that connect the surface with the subsoil, leaving the soil extremely susceptible to erosion. Old root channels and earthworm holes are eliminated as are the cracks between natural aggregates. The large pores, the very ones decreased in number by conventional tillage practices, are necessary to conduct water into the soil during rainstorms. The development of a "plow pan," a layer of compacted soil resulting from smearing action at the bottom of the plow, may retard both root penetration and water infiltration. Also, when natural aggregates break down, the soil surface crusts over more easily and leads to greater water runoff and erosion (see chapter 3). The smooth surface that develops leaves few "pockets" for temporary storage of water during intense storms, increasing the possibility of runoff and erosion. Runoff during rainstorms may also increase the

possibility of drought stress later in the season, because water that runs off the field does not infiltrate into the soil to become available to plants later on.

The increasing scale of agricultural operations, with larger equipment and fields, has increased the potential for soil erosion. In order to have larger and easier fields to work, farmers often remove protective hedgerows and terraces. This will eventually increase the intensity of runoff and erosion.

Reduced-Tillage Systems

Alternative tillage systems that create less potential for soil erosion than conventional tillage are becoming more popular. Many of these reduced-tillage systems save labor and machine use and/or lessen erosion. Because the soil is not turned over, these systems leave more surface residues, more residues in the first few inches of soil, and more surface roughness. Any system that leaves more than 30% of the soil surface covered with crop residues is considered to be a "conservation" tillage system by the U.S. Department of Agriculture's Soil Conservation Service. Reduced-tillage systems may require special planters that can cut through and push aside the surface residues and provide for good seed contact with the soil.

Reduced-tillage systems include no-till (or "zero tillage"), chisel plowing, ridge-till, and disking alone (without plowing). The effects of three of these systems, compared to the conventional moldboard plow plus disking, can be seen in figure 9.1. *Moldboard plowing and disking* (figure 9.1a) inverts the soil and buries crop residues and manures. The soil is left bare while many natural aggregates are broken down. Under *no-till planting* (figure 9.1b) only a narrow strip is disturbed to produce a seedbed while the residues on most of the soil surface remain undisturbed. If more soil disturbance is necessary to produce a good seedbed, a slightly wider strip of soil is worked by

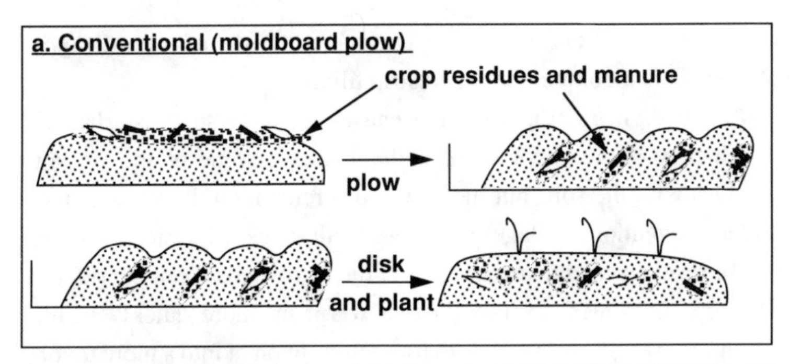

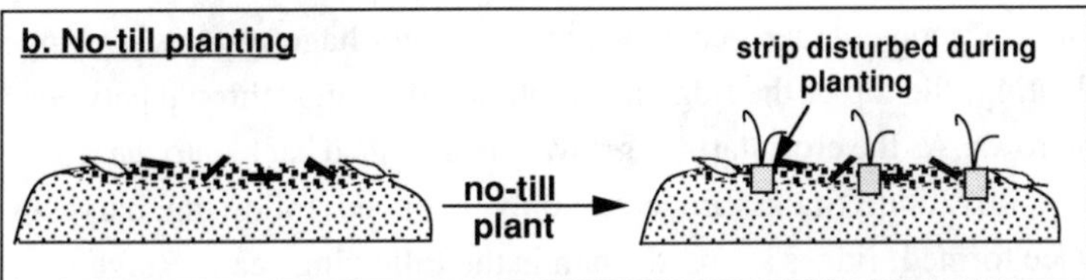

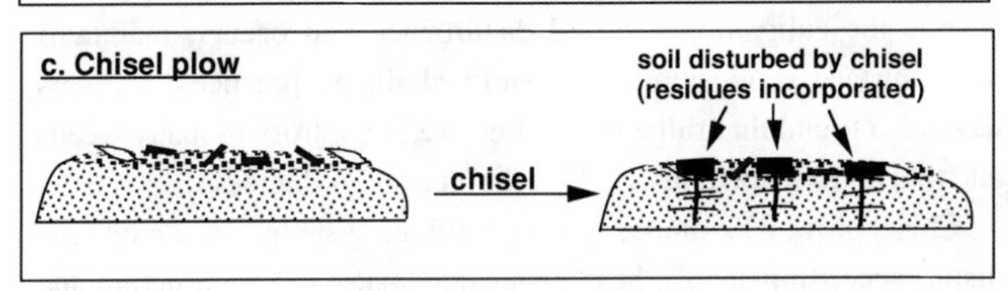

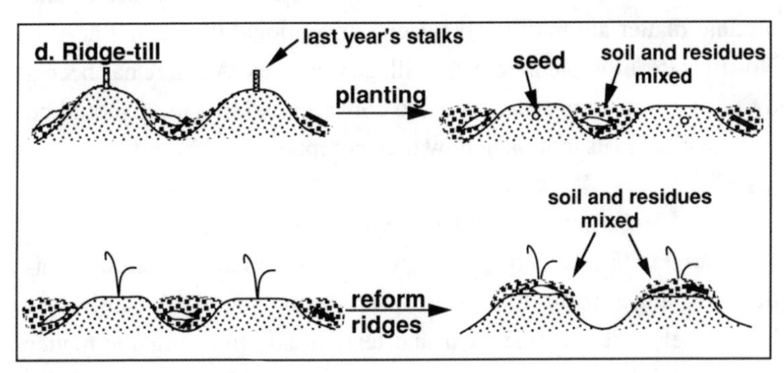

Figure 9.1. Effects of different tillage systems on residues and soils.

using fluted coulters in front of the opening disks of a no-till-type planter. This is sometimes called strip tillage.

Chisel plowing (Figure 9.1c) causes a lot of soil disturbance, which may be good for some soils. It mixes some of the surface residues with the soil, but more residues remain on the surface than under conventional tillage practices. With *ridge-till* systems (figure 9.1d), residues tend to accumulate in the furrows between the ridges. The tops of ridges dry rapidly and warm up more quickly in the spring, allowing farmers in nothern regions to plant into a more favorable environment than occurs with most other tillage systems. During planting, the top of the ridge is cut off and the soil is thrown between the rows. As the crop starts to grow, soil is pushed back onto the ridge by cultivation. Stalks are commonly chopped following harvest. Once formed, ridges are used again in the following years. Ridge-till, despite the cultivation and soil disturbance that occurs, maintains more surface residue than conventional tillage practices. Farmers have also found that ridge-till is effective in controlling many weeds and that it has helped them reduce their use of herbicides.

Chisel plow, disk-alone, and ridge-tillage systems cause moderate disturbance. In general, the changes they cause in soil structure and organic matter are midway between conventional tillage and no-till. No-till is the ultimate in reduced-tillage practices. As there has been a lot of research comparing no-till to conventional tillage, it should be worthwhile for us to look at how they compare.

Conventional and No-till Compared

Compared with conventional tillage practices, there are many potential advantages to no-till systems. These include significantly less labor, fuel, and machine wear and tear. In addition, organic matter can be maintained or increased more easily with no-till than with conventional practices. The positive influence of no-till on organic matter is due to reduced erosion of topsoil, and to the slower decomposition of surface residues as well as organic matter within the soil.

Numerous experiments performed under a variety of conditions clearly indicate that, compared to conventional tillage, there is much less soil erosion under no-till practices. Soils under no-till systems tend to have more natural aggregation, more large pores, and more surface residues. Earthworms are more plentiful under no-till practices and their channels are able to rapidly conduct water into the soil. Increased surface residues and roughness found in no-till soils can also slow down water moving across the surface. In addition, the surface residues behave as a mulch during the growing season, reducing evaporation from the soil.

No-till soils tend to have more moisture than conventionally tilled soils because of better water infiltration and less evaporation. This may be a disadvantage for poorly drained soils. It also means that leaching of various chemicals to the groundwater may be more extensive under no-till practices.

Organic matter decomposition takes place more slowly under reduced tillage, especially with no-till. Organic matter tends to accumulate on the surface of no-till soils. Numerous experiments show that residues on the surface decompose slower than when incorporated into the soil. In addition, organic matter within the soil also decomposes less rapidly when soils are not plowed. The result is that organic matter accumulates in soils after conversion to no-till.

There is a drawback to this accumulation of organic matter on and in soils under no-till. A lower rate of decomposition means the availability of nutrients, especially nitrogen, may be lower in the early stages of conversion to no-till. With time, this effect is not as important, since the increased amount of organic matter will result in a release of nitrogen more similar to that occurring with conventional tillage, even if the *percent* of nitrogen released from the organic matter is relatively low. In other words, the greater amount of organic matter present in no-till soils counteracts the lower percentage of each pound that will be converted to mineral forms.

No practice or system is without drawbacks or problems. And no practice is appropriate under all conditions. The main question must always be whether the advantages outweigh the disadvantages for specific situations. There have been reports of more problems controlling weeds under no-till than with conventional tillage. Most experiments have been performed under monocultures, usually corn, or very simple rotations. When you can't cultivate for weed control, more herbicides may be necessary. Herbicide tie-up by surface residues may contribute to the need for higher application rates. Perennial weeds take hold easier when soil is not cultivated annually and sometimes this makes for more pesticide use. However, by using vigorous cover crops and more creative rotations, including sod crops and small grains, controlling weeds under no-till conditions may not be as difficult as under monocultures or continuous row crops. Frequent plowing and cultivation brings weed seeds close to the surface where they germinate easily. But with no-till, new seeds are not brought up to the surface each year and there actually may be fewer weed seeds germinating after some years.

Sometimes slugs or plant diseases are encouraged by the high amount of surface residues common with no-till. The wetter soil conditions under no-till may also be a drawback for wet soils or in locations where leaching of agrichemicals to the groundwater may be an issue. In addition, soil under the mulch layer of no-till will be cooler in the spring than bare soil. This may cause planting to be delayed longer than under conventional tillage, a definite disadvantage in cool climates with short growing seasons. For northern soils, especially poorly drained ones, it might be best to use a reduced-tillage system other than no-till.

The nitrogen benefit of animal manures will usually be less when spread on the surface under no-till practice than when incorporated into the soil. Increased gaseous losses of ammonia from the surface application means that less nitrogen will be available to crops.

Summary

The traditional moldboard plow and disk-tillage system tends to cause rapid decomposition of soil organic matter and leave the soil susceptible to wind and water erosion. Reduced-tillage systems leave more surface residues, more intact large pores, and more soil aggregates. Water is able to infiltrate into the soil better with reduced tillage and this helps protect the soil from erosion. In addition, organic matter decomposes less rapidly under reduced-tillage systems. No-till systems have proven especially useful for maintaining and building up soil organic matter.

There are some potential drawbacks to no-till and other reduced-tillage systems. Soils are frequently cooler and wetter and lower in available nitrogen under no-till than with conventional tillage. However, no-till, ridge-till, chisel plowing, and disking without plowing are all systems which have proven useful in large areas of the country.

Sources

Allmaras, R. R., G. W. Langdale, P. W. Unger, R. H. Dowdy, and D. M. Van Doren. 1991. Adoption of conservation tillage and associated planting systems. In *Soil management for sustainability,* ed. R. Lal and F. J. Pierce, pp. 53–83. Ankeny, Iowa: Soil and Water Conservation Society.

Dick, W. A., D. M. Van Doren, Jr., G. B. Triplett, Jr., and J. E. Henry. 1986. *Influence of long-term tillage and rotation combinations on crop yields and selected soil parameters.* I. Results obtained for a mollic ochraqualf soil. Research Bulletin 1180. The Ohio State University, Wooster, Ohio.

Follett, R. H., S. C. Gupta, and P. G. Hunt. 1987. Conservation practices: Relation to the management of plant nutrients for crop production. In *Soil fertility and organic matter as critical components of production systems,* ed. R. F. Follett, J. W. B. Stewart, and C. V. Cole, pp. 19–52. Madison, Wis.: American Society of Agronomy.

10

• • •

Composts

The reason of our thus treating composts of various soils and substances, is not only to dulcify, sweeten, and free them from the noxious qualities they otherwise retain . . . [Before composting, they are] apter to ingender vermin, weeds, and fungous . . . than to produce wholsome plants, fruits and roots, fit for the table.—J. Evelyn, seventeenth century

Composting farm wastes as well as organic residues from off the farm has become a widespread practice. Accepting and composting lawn and garden wastes can provide some income for farmers near cities and towns. They can charge for accepting the wastes and for selling compost. Some farmers, especially those without animals or sod crops, may want to utilize the compost as a source of organic matter for their own soils.

Decomposition of organic materials takes place naturally in forests and fields all around us. Composting is the art and science of combining available organic wastes so that they decompose to form a uniform and stable finished product. This reduces bulk, stabilizes soluble nutrients, and hastens the formation of humus. Most organic materials, such as manures, crop residues, grass clippings, leaves, sawdust, many kitchen wastes, and sewage sludges, can be composted. Composts are excellent organic amendments for soils.

The composting process involves rapid decomposition and stabilization of organic residues. The microorganisms that do much of the work need high temperatures, plenty of oxygen, and moisture. These heat-loving, or *thermophilic,* organisms work best between about 110° and 130°F. Temperatures above 140°F can develop in compost piles, but this usually slows down the process. At temperatures below 110°F the less active types of organisms take over and the rate of composting again slows down. The composting process is slowed down by anything that inhibits good aeration or the maintenance of high enough temperatures. Organic matter decomposition can occur at low pile temperatures, but the process isn't as rapid or as complete as with high-temperature composting.

Rules for Making Composts

The organic materials used should have plentiful carbon and nitrogen available for microorganisms to use. High nitrogen materials such as chicken manure can be mixed with high carbon materials like hay, straw, leaves, or sawdust. This is sometimes done by building the compost pile out of alternate layers of high-carbon and high-nitrogen materials. Turning the pile mixes the materials together. Manure mixed with sawdust or wood chips used for bedding can be composted as is. Composting can occur most easily if the average C:N ratio of the materials is about 25 to 40 parts carbon for every part nitrogen.

Avoid using certain materials such as coal ash, wood chips from pressure-treated lumber, manure from pets, and large quantities of fats, oils, and waxes. These types of materials are either difficult to compost or may result in compost containing chemicals that can harm crops.

Consider the size of the materials used in the compost pile. The materials need to fit together in a way that allows oxygen from the air to flow in freely. But it is also important that not too much heat escape from the center of the pile. If small-sized particles are used, a "bulking agent" may be needed to make sure that enough air can enter the pile. Sawdust, dry leaves, hay, and wood shavings are frequently used as bulking agents. Tree branches will need to be "chipped" and hay chopped up so that it doesn't mat and slow down composting. Composting will take longer when large particles are used, especially those resistant to decay.

The pile needs to be large enought to retain much of the heat that develops during composting, but not so large and compacted that air can't easily flow in from the outside. Compost piles should be 3 to 5 feet tall and about 6 to 10 feet across the base *after* the ingredients have settled (see figure 10.1). Easily condensed material should initially be piled higher than 5 feet. It is possible to have long windrows of composting materials as long as they are not too tall or wide.

The amount of moisture in the pile is important. If the materials mat and rain water can't drain easily through the pile, it may not stay aerobic in a humid climatic zone. On the other hand, if composting is done inside a barn or under dry climatic conditions, the pile may not be moist enough to allow composting to occur. It may be necessary to add water to the pile. In fact, even in a humid region it is a good idea to moisten the pile at first, if mainly dry materials are used. However, if something like liquid manure is used to provide a high-nitrogen material, sufficient moisture will most likely be present to start the composting process. The ideal moisture content of compost is about 40 to 60%, or about as damp as a wrung-out sponge.

Turning the compost pile speeds up decomposition. It exposes all the materials to the high-temperature conditions at the center of the pile. At any given time, the materials at the top and on the sides of the

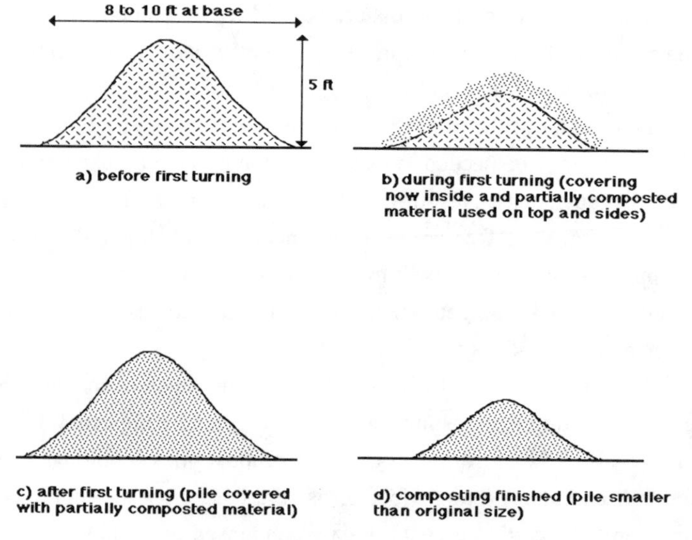

Figure 10.1. Compost-pile dimensions and turning techniques.

pile are barely composting. But they do provide insulation for the rest of the pile. Turning the pile rearranges all the materials and creates a new center. If piles are turned every time the interior reaches and stabilizes at about 140°F for a few days, it is possible to complete the composting process within months. But, if you only turn the pile occasionally, it may take a year or longer to complete the process. There is now equipment available to quickly turn long compost windrows at large-scale composting facilities. There are also tractor-powered compost turners designed for composting on farms.

Following high temperature composting the pile should be left to cure for about one to three months. The importance of curing is greater if the active (hot) process is shortened or poorly managed. There is no need to turn the pile at this time because you are not trying to stimulate maximum decomposition and there is less need for rapid oxygen entry into the pile's center when decomposition rate is slow. Curing the pile furthers aerobic decomposition of resistant chemicals and larger particles. Also, common soil organisms populate the pile during curing, the pH becomes closer to neutral, and ammonium is converted to nitrate. It is thought that the processes that occur during curing give compost some of its disease-suppressing qualities.

Using Composts

Finished composts are *not* good sources of readily available nutrients. During composting nutrients are converted into more stable forms. But composts can greatly help the fertility of your soil by increasing organic matter and by slowly releasing nutrients. Composts can be used on turf, in flower gardens, and for vegetable and agronomic crops. Composts can be spread and left on the surface or incorporated into the soil by plowing or rototilling. Composts are also used to grow greenhouse crops and as the basis of some potting-soil mixes.

Advantages of Composting

Composted material is less bulky than the original material, and easier and more pleasant to work with. During the composting process carbon dioxide and water are lost to the atmosphere and the size of the pile decreases by 30 to 60%. In addition, many weed seeds as well as disease-causing organisms may be killed by the high temperatures in the pile. Unpleasant odors are eliminated. Flies, a common problem around manures and other organic wastes, are much less of a problem with composts. Composting reduces or eliminates the decrease in nitrogen availability to crops (immobilization) that commonly occurs when materials such as sawdust or straw are added directly to soil. Composting is also very useful for recycling kitchen wastes, leftover crop residues, weeds, and manures. Many types of local organic waste, such as apple pumice, lake weeds, leaves, and grass clippings, are usefully composted.

There is some evidence that compost application lowers the populations of some plant-disease organisms. This may help explain some of the broad benefits to plant growth that have been attributed to well-cured compost.

If you have a lot of organic waste to dispose of and not much land, composting may be very helpful. Also, since making compost decreases the solubility of nutrients, composting may help lessen pollution in streams, lakes, and groundwater. On many poultry farms and on beef feedlots, where high animal populations on limited land may make manure application a potential environmental problem, composting may be the best method for handling the wastes.

Without denying these good reasons to compost, there are frequently very good reasons to just add organic materials directly to the soil without composting. Compared to fresh residues, composts may not stimulate as much production of the sticky gums that help hold aggregates together. Also, some uncomposted materials have more nutrients that are readily available to feed plants than do composts. If your soil is very deficient in fertility, plants may need readily available nutrients from residues. Finally, more labor and energy are usually needed to compost residues before applying than to simply apply the uncomposted residues directly.

Summary

Composting is the rapid, high-temperature decomposition of organic residues carried out mainly by heat-loving microorganisms. Composting requires good aeration of the pile as well as adequate moisture content. Finished compost is pretty stable and when added to soil provides organic matter as well as a slow-release source of nutrients. Because of its consistency and other qualities, composts can be utilized in almost every situation in which plants are grown, including on turf, home gardens, in potting-soil mixes, and on agricultural soils. There are a number of drawbacks to composting, but the potential benefits are numerous. Farmers near urban and suburban sources of organic waste may find composting an income-generating practice.

Sources

Martin, D. L., and G. Gershuny, eds. 1992. *The Rodale book of composting: Easy methods for every gardener.* Emmaus, Penn.: Rodale Press.

Rothenberger, R. R., and P. L. Sell. n.d. *Making and using compost.* University of Missouri Extension Leaflet (File: Hort 7 2/76/20M). Columbia, Missouri.

Rynk, R., ed. 1992. *On-farm composting.* NRAES-54. Ithaca, N.Y.: Northeast Regional Agricultural Engineering Service.

Staff of *Compost Science,* eds. 1981. *Composting: Theory and practice for city, industry, and farm.* Emmaus, Penn.: The JG Press.

11

• • •

Decreasing Soil Erosion

The dust storm hit and it hit like thunder.
It dusted us over, it dusted us under . . .
So long! It's been good to know you.
So long! It's been good to know you.
So long! It's been good to know you.
This dusty old dust is a gettin' my home.
And I've got to be drifting along.©
—Woody Guthrie, 1940

The dust storms that hit the center of the U.S. during the 1930s were responsible for one of the great migrations in our history. As Woody Guthrie pointed out in his songs, conditions were so bad that people had no alternative to abandoning their farms. They moved to other parts of the country in search of work. Although changed climatic conditions and agricultural practices improved the situation for a time, there was another increase in wind and water erosion during the 1970s and 1980s.

Erosion of soil by wind and water has been going on since the beginning of time. We should expect some erosion to occur on almost all soils. But erosion usually increases when we use soils for agriculture. Erosion is more of a problem on some soils and in some regions than in others. However, it is the major hazard or limitation to the use of about one-half of all cropland in the United States! And on much of this land erosion is occurring fast enough to reduce future productivity. As we discussed earlier, erosion is an organic matter problem because it usually removes some of the soil layer highest in organic matter, the topsoil.

It's all right to have a small amount of erosion as long as new topsoil can be created as rapidly as soil is lost. The amount of soil that you can lose each year while still maintaining reasonable productivity is called the *soil loss tolerance* or *T value*. If you have a deep soil with a rooting depth of greater than 5 feet, T is 5 tons per acre each year. While this sounds like a lot of soil loss, keep in mind that the weight of an acre of soil to 6 inches is about 2 million pounds, or 1,000 tons. So 5 tons is equivalent to about 3 hundredths of an inch ([5/1,000] x 6 inches = 0.03 inch). If soil loss continues at this rate, at the end of 33 years about 1 inch will be lost. But on deep soils with good management of organic matter the rate of topsoil creation can balance this loss.

The soil loss tolerance amount is gradually reduced for soils with less rooting depth. When the rooting depth is shallower than 10 inches, T is about 1 ton per acre each year. This is the same as 0.006 inch per year and is equivalent to 1 inch of loss in 167 years.

When your soil loss is higher than the tolerance value, productivity will suffer. Yearly losses of 10 or 15 tons or more per acre occur in many fields. But there are practices you can use to help reduce runoff and soil losses. For example, researchers in Washington State found that erosion on winter wheat fields was about 4 tons each year when a sod crop was included in the rotation compared to about 15 tons when sod was not included. Another example is from an Ohio experiment where runoff from conventionally tilled and no-till continuous-corn fields were monitored. Over a four-year period, runoff averaged about 7 inches of water each year for conventional tillage and less than one tenth (0.1) inch for the no-till planting system.

Erosion reduction works by either decreasing the amount of water flowing over the soil or by keeping soil in such a condition that it can't erode easily. Many conservation practices actually have both effects. The soil organic matter management practices we discussed in the earlier chapters all influence erosion. We'll also briefly cover other important practices for keeping erosion to a minimum.

Adding Organic Materials

When you maintain good soil organic matter levels, you help keep topsoil in place. A soil with more organic matter usually has better tilth and less surface crusting. This means that more water is able to infiltrate into the soil instead of running off the field. When less water runs off the field, there is less soil erosion. So, when you build up organic matter by adding manures, composts, and other organic residues you help control erosion by making it easier for rainfall to enter the soil.

Organic materials added regularly to soils also result in larger and more stable soil aggregates. The larger aggregates are not eroded by wind or water as easily as smaller ones. And don't forget the effects of many residues on soil organisms. Surface residues provide food for large numbers of earthworms that will go to work for you. Earthworm channels conduct water quickly into the soil, lessening the amount of water running off the field. Many more details about using crop residues, manures, and composts are given in chapters 5, 6, and 10.

Cover Crops

Growing cover crops decreases erosion in a number of ways. The cover crop adds organic residues to the soil and in this way helps to maintain organic matter levels. Cover crops frequently grow during seasons when the soil is especially susceptible to erosion. Their roots help to bind soil and hold it in place. Also, raindrops lose most of their energy when they hit leaves and drip to the ground. There's less soil crusting because the rain can't break apart as many aggregates when the surface is covered with plants. These effects help explain why cover crops have been found to increase water infiltration and reduce erosion. For more information about cover crops, see chapter 7.

Reduced Tillage

Reduced-tillage practices result in less soil disturbance and leave significant quantities of residue on the surface. Any system that leaves more than 30% of the soil surface covered with residue is considered soil conserving. Surface residues are important because they intercept raindrops and lessen their force. Residues can also slow down water running over the surface. The amount of residue on the surface may be close to zero for the conventional moldboard plow and disk tillage system. On the other hand, no-till planting commonly leaves 90% or more of the surface covered with residues. Other reduced-tillage systems, such as chiseling and disking (as a primary tillage operation), typically leave more than 30% of the surface covered as long as the crop leaves a lot of residues.

Reduced tillage usually leaves the soil surface a little rough, with small depressions scattered around the fields. These small depressions store water temporarily during rainstorms. The water in these puddles can then slowly infiltrate into the soil instead of running off the field with sediments. No-till does the best job of conserving topsoil and organic matter. More details on tillage are discussed in chapter 9.

Sod Crops in Rotation

Grass and legume sod crops can help lessen erosion. Sod crops have a beneficial effect because they maintain a cover on most of the soil surface for the whole year. Their extensive root systems hold soil in place. Permanent sod is a very good choice for very steep soils, or other soils that erode easily. Details on the importance of growing sod crops as part of a crop rotation are discussed in chapter 8.

Other Soil-Conservation Practices

Diversion ditches are frequently very helpful in channeling water so that it can leave a field easily without flowing over the entire area. Grassed waterways for diversion ditches and other field water channels trap sediments and keep them on the field. Tilling and planting along the contour is a simple practice that helps control erosion. When you work along the contour instead of up and down slope the wheel tracks and depressions caused by your plow, harrow, or planter will move water across the face of the hill to a grass waterway or other channel. Strip-cropping along the contour, with alternating strips of row crops and sod crops, can allow you to grow row crops on many sloping soils without too much erosion. Terracing soil in hilly regions is an expensive practice, but one which results in level soil that is much less likely to erode. Well-constructed and maintained structures can last for a long time, frequently making it worthwhile to construct terraces despite the high initial investment.

Summary

Many conservation practices can help you reduce erosion to acceptable levels. If you're intent on maintaining and building up soil organic matter you must keep erosion in check. But promoting soil organic matter is in itself important to controlling erosion!

There are conservation practices that are aimed specifically at lessening soil erosion. Soil management for the best control of erosion requires that you use a combination of some of the practices discussed in this chapter.

Sources

American Society of Agricultural Engineers. 1985. *Erosion and soil productivity.* Proceedings of the national symposium on erosion and soil productivity, December 10–11, 1984, New Orleans, Louisiana. American Society of Agricultural Engineers Publication 8–85. St. Joseph, Michigan.

Edwards, W. M. 1992. Soil structure: Processes and management. In *Soil management for sustainability,* ed. R. Lal and F. J. Pierce, pp. 7–14. Ankeny, Iowa: Soil and Water Conservation Society. This is the reference for the Ohio experiment on the monitoring of runoff.

Lal, R., and F. J. Pierce, eds. 1991. *Soil management for sustainability.* Ankeny, Iowa: Soil and Water Conservation Society.

Reganold, J. P., L. F. Elliott, and Y. L. Unger. 1987. Long-term effects of organic and conventional farming on soil erosion. *Nature* 330:370–72. This is the reference for the Washington State study of erosion.

Soil and Water Conservation Society. 1991. Crop residue management for conservation. Proceedings of a national conference, August 8–9, 1991, Lexington, Kentucky. Ankeny, Iowa: Soil and Water Conservation Society.

United States Department of Agriculture. 1989. *The second RCA appraisal: Soil water, and related resources on nonfederal land in the United States, analysis of conditions and trends*. Washington, D.C.: Government Printing Office.

12

• • •

Putting It All Together

Soil . . . under careful and wise handling . . . is able to maintain its productiveness with scarcely a trace of diminution for decade after decade and century after century.—E. Fippin, 1911

In the earlier chapters of part 2 we've gone over many different ways to manage soils, crops, and residues in order to build up and maintain good levels of organic matter. We looked at each one as a separate strategy: animal manures, cover crops, rotations, composts, reduced tillage, and erosion control. In the real world, we need to combine a number of these approaches and use them together. In this chapter we'll briefly review the practices that maintain or increase soil organic matter. If the thought of making a lot of changes on your farm is overwhelming, you can start with only one or two practices that promote soil organic matter buildup. Not all of these suggestions are meant to be used on every farm.

If at all possible, you should use rotations that utilize grass, legume, or a combination of grass and legume sod crops, or crops with large amounts of residue as important parts of the system. Leave residues from annual crops in the field or, if removed for composting or to use as bedding for animals, return them to the soil as manure or compost. Use cover crops when soils would otherwise be bare during a period of the year both to add organic matter and to decrease erosion. Cover crops will also help maintain soil organic matter in resource-scarce regions where there are no possible substitutes to using crop residues for fuel or building materials.

Raising animals or having access to animal wastes from nearby farms gives you a wider choice of economically sound rotations. Rotations that include sods make hay or pasture available for use by dairy and beef cows, sheep, and goats. In addition, on mixed crop-livestock farms, animal manures can be applied to cropland. It's easier to maintain organic matter on a diversified crop-and-livestock farm where sod crops can be fed to animals and manures returned to the soil. However, growing crops with high quantities of residues plus frequent use of green manures and composts from vegetative residues will help maintain soil organic matter even without animals.

You can maintain or increase soil organic matter more easily when you use reduced-tillage systems, especially no-till, instead of the conventional moldboard plow and disk system. The decreased soil disturbance under reduced tillage slows the rate of organic matter decomposition and helps to maintain a soil structure that allows rainfall to infiltrate rapidly. Leaving residue on the surface encourages the development of high earthworm populations, which also improves soil structure. Compared to conventional tillage, soil erosion is greatly reduced under minimum-tillage systems and this helps to keep the organic matter in the rich topsoil. Any practice that reduces soil erosion, such as contour tillage, strip-cropping along the contours, and terracing, also helps maintain soil organic matter.

Even if you use minimum-tillage systems to leave significant quantities of residues on the surface and decrease the severity of erosion, you should also be using crop rotations. In fact, it may be more important to rotate crops when a lot of residues remain on the surface. The decomposing residues can harbor many insect and disease organisms. These problems can be worse in monoculture with no-till practices than with conventional tillage.

Organic matter is the key to soil fertility, but it is not the only thing to consider. Don't give up your recognition of the importance of other practices just because your soil has a high level of organic matter. For example, it's not good to work soils that are wet enough to lead to excess compaction. And don't forget to pay attention to nutrient availability. If you do a better job of building up and maintaining organic matter you will probably not need to apply lime or commercial fertilizer as frequently as before. While you can supply a lot of nitrogen with manures and legumes, you still may need to use potassium or phosphorus fertilizers. In some situations, other nutrients may also be needed. In the drier parts of the country, excess sodium or salts may cause problems that require remedies. Good levels of soil organic matter will make it easier to improve soils with these problems, but adding gypsum or leaching salts may also be necessary.

Test your soils regularly and apply lime and fertilizers when they are needed. Testing soils every year or two on each field is one of the best investments you can make. If you keep the report forms, or record the results, you will be able to follow the fertility changes over the years. Seeing soil-test changes will help you fine-tune your practices. Soil-testing laboratories usually charge extra for an organic matter determination. It's probably worth the money every few years just to track changes.

A word of caution when comparing your soil test organic matter levels with those discussed in this book. If your laboratory reports organic matter as "weight loss" at high temperature the numbers may be higher than if the lab had used the traditional wet chemistry method. A soil with 3% organic matter by wet chemistry might have a weight-loss value of between 4 and 5%. You can use either method to follow changes in your soil, but when you compare soil organic matter of samples run in different laboratories, make sure the same methods were used.

When you creatively combine a number of practices that promote the buildup and maintenance of organic matter, then most of your farm's soil-fertility problems will have been solved along the way. The health and yield of your crops should improve. The soil will have more available nutrients, more water for plants to use, and better tilth. There should be fewer problems with diseases, nematodes, and insects.

By concentrating on the soil-conserving practices that build up organic matter you will also leave a legacy of land stewardship for your children and their children to inherit and follow.

Part Three: Dynamics and Chemistry

13

• • •

Organic Matter Dynamics and Nutrient Availability

Organic matter functions mainly as it is decayed and destroyed. Its value lies in its dynamic nature.—W. A. Albrecht, 1938

Influences on Soil Organic Matter Levels

A wide variety of environmental, soil, and agronomic influences determines the organic matter level in a particular soil. Let's briefly review some of these influences.

Temperature. Soils in cooler climates are usually higher in organic matter because of slower mineralization (decomposition) rates.

Moisture. Soil organic matter levels are usually higher as mean annual precipitation increases. This is probably because greater plant growth leaves more residues, while poor aeration in moist soils reduces the rate of mineralization.

Soil texture. Soil organic matter tends to increase as the percentage of clay content increases, probably due mainly to bonds between clay and organic matter that protect organic molecules from microbial attack.

Soil structure and drainage class. Poorly drained soils tend to be higher in organic matter mainly because poor aeration decreases mineralization rates.

Tillage practices. The more intensively a soil is tilled, the more rapidly soil organic matter is mineralized or lost through erosion.

Rotation practices. The more frequent the use of sod-type legume and grass crops in a rotation, the more organic matter. This is a result of regular additions of high amounts of root residues, low organic matter decomposition rates in unplowed soils, and reduced erosion losses of organic matter–rich topsoil.

Manure use. Higher manure-application rates result in higher organic matter levels.

Patterns of Change in Soil Organic Matter Levels

There are four main patterns of soil organic matter change that result from crop, soil, and manure management practices (figure 13.1). Soil organic matter can increase and then come to a plateau (A), stay pretty much the same (B), fluctuate down and up—or up and down (C), or decrease rapidly and then come to a plateau where there is not much year-to-year change (D).

In scenario A, soil organic matter gradually increases and then reaches a plateau, after which there is little increase. This pattern occurs when the amount of residues added each year is greater than the yearly amount of soil organic matter decomposition and loss through erosion. This can happen with any cropping system if enough external residues, such as manures and composts, are applied to the field. Organic matter will increase under the following circumstances: (1) application of more than 20 to 30 tons of beef or dairy manure per acre each year, or other manures with equivalent solids content; (2) growth of a crop that produces high quantities of residues that are left in

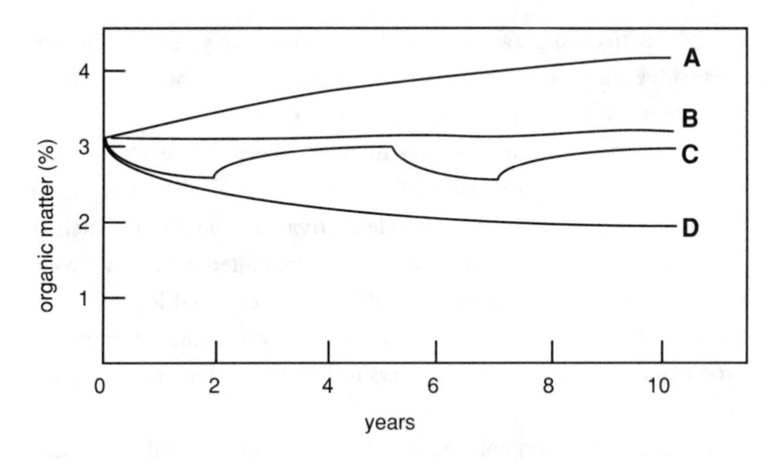

Figure 13.1. Examples of common patterns of change in soil organic matter resulting from crop, tillage, and manure management. In (a) gains exceed losses, in (b) gains equal losses, in (c) cyclical fluctuations occur, and in (d) rapid decrease leads to a lower plateau.

the field and application of smaller amounts of manures; (3) seeding down of a field to sod-type grass or legume crops after years of growing cultivated row crops; or (4) changeover from conventional tillage practices to no-till planting.

In scenario B, soil organic matter remains unchanged, as annual additions just balance losses. This might occur in the following circumstances: (1) use of conventional tillage practices to grow a crop that leaves few residues and application of about 20 to 30 tons of dairy manure per acre each year, or other manures with the equivalent solids content; (2) use of conventional tillage practices to grow a crop that produces large amounts of residues that are left in the field—around 2 to 4 tons dry weight per acre per year; or (3) use of conventional tillage practices to grow a crop with only moderate yields and moderate residue additions with application of about 10 tons of dairy manure per acre per year or regular green manure crops.

In scenario C, organic matter fluctuates between decreasing and increasing levels in a regular, cyclical fashion. This will mainly occur when you use rotations that include cultivated annual crops such as corn silage, soybeans, cotton, and wheat that alternate every two to five years with sod crops such as alfalfa, clovers, and legume-grass mixes. Organic matter levels will decrease while annual crops are grown and then increase in the years following seeding down to a sod crop.

In scenario D, organic matter decreases rapidly and eventually reaches a lower plateau. This happens when crops are grown on soils that have previously been in forest, native grassland, or long-term hay. The most severe form of this pattern occurs when row crops are raised every year using the moldboard plow as the primary tillage tool, without the addition of manures or green manures. Although high yielding crops help by leaving high amounts of residues, there will usually still be an organic matter decline from the original levels. However, the lower plateau, or equilibrium level, will stabilize sooner when crop yields and residue additions are high.

Mathematical Approaches to Organic Matter Dynamics

The understanding of soil organic matter dynamics, the buildup, losses, and steady state situations, can be supported by mathematical approaches. A few different approaches are introduced below.

A Simplified Equilibrium Model

The amount of organic matter in soils is a result of the balance between the gains and losses of organic materials. Let's use the abbreviation *om* as shorthand for *organic matter*. Then the change in soil organic matter during one year (*om change*) can be represented as follows:

$$om\ change = gains - losses \quad [1]$$

If gains are greater than losses, organic matter accumulates and the term *om change* is positive. When gains are less than losses, organic matter decreases and *om change* is negative. Remember that *gains* refer not to the amount of residues added to the soil each year but rather to the amount of residue remaining at the end of the year. This is the fraction (f) of the fresh residues added that is not decomposed during the year multiplied by the amount of fresh residues added (A), or $gains = (f)(A)$.

If you follow the same cropping and residue/manure addition pattern for a long time, a steady-state situation usually develops in which gains and losses are the same and *om change* = 0. Losses consist of the percentage of organic matter that's mineralized, or decomposed, in a given year (let's call that percentage k) multiplied by the amount of organic matter (*om*). Another way of writing that is $Losses = k(om)$. The amount of organic matter that will remain in a soil under steady-state conditions can then be estimated as follows:

$$om\ change = 0 = gains - k\,(om) \quad [2]$$

Because in steady-state situations *gains* = *losses,* then $gains = k\,(om)$, or

$$om = gains/k \quad [3]$$

A large increase in soil organic matter can occur when you supply very high rates of crop residues, manures, and composts or grow cover crops on soils in which organic matter has a very low rate of decomposition (k). Under steady-state conditions, the effects of residue addition and the rate of mineralization can be calculated using

Table 13.1. Estimated Levels of Soil Organic Matter (Steady-State Conditions)

	k (Mineralization Rate)				
Residue Additions (lb/a/yr)	*0.01*	*0.02*	*0.03*	*0.04*	*0.05*
	% Organic Matter				
500	2.5	1.3	0.8	0.6	0.5
1,000	5.0	2.5	1.7	1.3	1.0
1,500	7.5	3.8	2.5	1.9	1.5
2,000	10.0	5.0	3.3	2.5	2.0

equation [3]. The results of these calculations, assuming 1 acre to 6-inch depth weighs 2 million pounds, can be seen in table 13.1.

A cultivated clay soil with reasonably good drainage may have a rate of organic matter decomposition of 2% each year ($k = .02$). With a high rate of residue addition, 2,000 pounds dry weight per acre per year, this soil can have a 5% organic matter level at equilibrium. However, for a very well drained sandy-loam soil with a 4% rate of organic matter decomposition per year the same residue-application rate will result in only 2.5% organic matter.

It is possible for different conditions to result in the same percentage of soil organic matter. For example, a soil with 2.5% organic matter can result from a low rate of addition (500 lb/a/yr) and a low mineralization rate of organic matter (1%). But it is also possible to obtain 2.5% organic matter as a result of a higher addition rate (1,500 lb/a/yr) and a higher rate of mineralization (3%).

AN EXPONENTIAL MODEL

Another model is used to describe changes in organic matter levels after residues are added to soils or under particular crop or manure management systems that continue for a long period of time. You should be able to understand the description of the model even without knowing calculus. This dynamic model is quite similar to the simplified equilibrium model just discussed. But with this exponential model it is possible to evaluate practices for any time period, even before equilibrium has been reached and annual gains are equal to losses.

Equation [2], $om\ change = 0 = gains - k\,(om)$, can be rewritten as follows:

$$\mathrm{d}(om)/\mathrm{d}t = gains - k\,(om) \qquad [4]$$

When this equation is integrated and k is assumed to be such that $k(om) = gains$, the following is obtained

$$om = om_{eq} - (om_{eq} - om_0)e^{-kt} \qquad [5]$$

where om = the organic matter content at time t, om_0 = the initial organic matter content, om_{eq} = the equilibrium organic matter content, and t = years since practice begun.

At the very start, time is zero and $e^{-kt} = e^0 = 1$. In this case the organic matter content is the same as the initial level, $om = om_0$. Now the expression e^{-kt} is the same as $1/e^{kt}$. For long periods of time t is very large and, because 1 divided by a large number is close to zero, e^{-kt} approaches 0. When this happens, om gets closer to om_{eq}. Since om_{eq} is also $gains/k$, equation [5] becomes the same as equation [3] when t is large [$om = om_{eq} = gains/k$].

There are even more sophisticated models for representing changes in organic matter levels. These take into account the fact that different

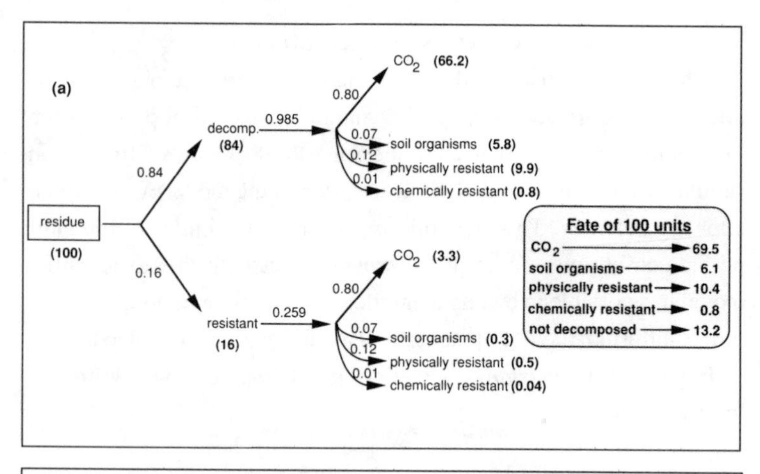

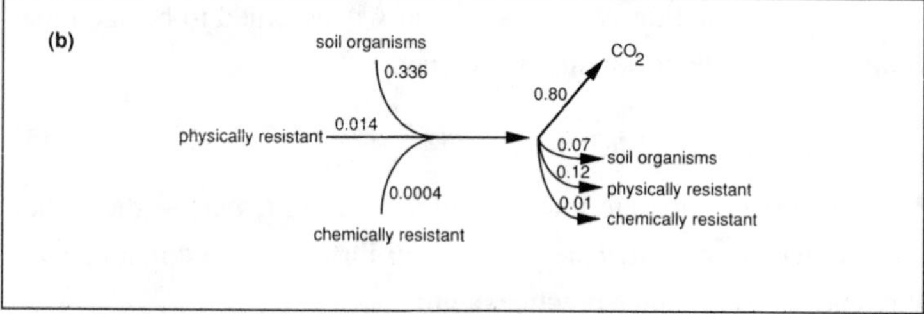

Figure 13.2. Multicomponent model for organic matter decomposition in the first year including (a) recently applied residues, and (b) older organic matter fractions. The coefficients were calculated according to equation [5] using data from a variety of experiments in England. Based on Jenkinson and Rayner, 1977.

parts of soil organic matter and surface residues will mineralize at different rates. Sugars, proteins, and starches decompose rapidly, while cellulose decomposes a little more slowly, and lignin even more slowly. Organic matter molecules bonded to a clay particle will decompose more slowly than the same organic molecules free in the soil.

In figure 13.2 the estimated numerical decomposition rates are presented for various fractions of organic matter. In this case, 84% of the added residue is considered readily decomposable, but only 82.7% (98.5% of the 84%) actually decomposes in the first year. Of the amount that actually decomposes, 80% ends up as CO_2 due to microbial respiration. Of the 20% that remains in the soil at the end of a year, 7% ends up as part of the soil organisms, 12% is physically stabilized and 1% is chemically stabilized. Only 25.9% of the resistant parts of the residue actually decomposes in the first year (lower line in figure 13.2a). However, the part that does decompose has the same fate as discussed above, with 80% ending up as CO_2, 7% as organisms, and so forth.

The fate of 100 units of carbon in fresh residues added to the soil is given in the parentheses in Figure 13.2a and a summary appears in the box on the right. By far the largest amount, about 70%, is lost as CO_2 while about 13% is not yet decomposed.

Of the organic matter already present in the soil, 33.6% of the carbon in soil organisms, 1.4% of the physically stabilized organic matter, and 0.04% of the chemically stabilized organic matter reacts during the year (see figure 13.2b). The fate of the carbon that reacts is assumed to be the same as discussed above; 80% of the amount that reacts ends up as CO_2, 7% as soil organisms, and so on.

Nutrient Availability

As soil organic matter is decomposed, nutrients are converted into mineral forms that are available to plants. Although numerous essential elements are made available by the mineralization of organic matter, we will focus on nitrogen. This element is needed in very large amounts: plants may take up from 100 to more than 300 pounds of nitrogen per acre. Nitrogen is also the nutrient most likely to be deficient in agricultural soils. For these reasons, plus the fact that there is more information on soil nitrogen dynamics than on other nutrients, it makes sense to concentrate on this element.

The nitrogen that is readily available to plants is almost completely present in soils in the form of inorganic chemicals. In reasonably aerated soils, nitrate (NO_3^-) will be the main source of available nitrogen for plants to use. There are a number of sources for this inorganic nitrate molecule. It can be mineralized from soil organic matter and organic residues or manures. It also can be added directly as a component of a synthetic fertilizer, such as ammonium nitrate. Ammonium and urea in commercial fertilizers are also converted to nitrate by chemical and biological reactions.

A high percentage of the nitrogen needs of crops growing on fertile soils may come from decomposing soil organic matter in the field. This has caused many people to believe that soil organic matter, or some other factor that is closely related to it, can serve as an indication of nitrogen availability. However, the reason organic matter is not very useful as a predictor of nitrogen-supplying capability is that two soils with the same percentage of organic matter may have different mineralization rates. Since organic matter is about 5% nitrogen, a soil with 2.5% organic matter will have 2,500 pounds of N per acre in the first 6 inches (or $0.025 \times 0.05 \times 2{,}000{,}000$ lb of soil per acre to 6 in. depth). If the rate of mineralization (*k*) is 1%, 25 pounds of nitrogen will be mineralized per year (0.01 x 2,500). A soil with the same amount of organic matter, but with a 3% rate of mineralization, will produce about 75 pounds of available N each year (0.03 x 2,500).

Mineralization Rates

The percentage of organic matter that mineralizes each year (*k*) tends to be less in soils with higher organic matter levels than in those with

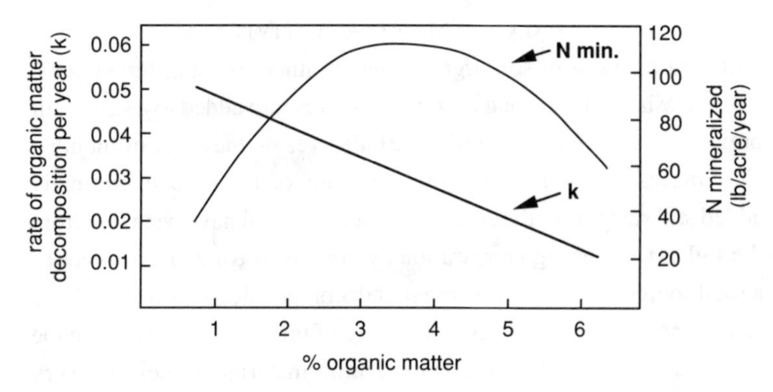

Figure 13.3. Nitrogen availability and its relation to percent organic matter and decomposition rates. Redrawn from F. R. Magdoff, 1991, p. 1512. Courtesy of Marcel Dekker, Inc.

lower levels. This occurs because soils in a given region have different organic matter contents mainly because of different rates of mineralization. Soils with high clay contents or restricted drainage tend to retain organic matter, because the organisms can't decompose the organic matter easily. Well-drained sands and loamy sands usually support very high rates of decomposition of any added organic materials and are therefore low in organic matter. Consequently, the amount of nitrogen that is made available to plants by mineralization of organic matter ($k \times$ *organic-N*) has a parabolic relationship with organic matter, or organic-N, content (see figure 13.3). Soils low in organic matter have low amounts of mineralized N, despite their high k, because they don't have much organic-N to mineralize. Soils high in organic matter have relatively low amounts of mineralized N because of the low k. Thus, organic matter or organic-N is not a very accurate predictor of the quantity of N that will be mineralized in the field.

Organic Matter Activity

Different fractions of soil organic matter mineralize at different rates. When fresh residues, such as green manures, are added to a soil, their organic-C and organic-N will have half-lives on the order of months. This means that within months only half of the carbon originally added still remains in the soil. The other 50% will have been lost from the soil as CO_2 through respiration by organisms using the residues as a food source. After this first burst of decomposition is completed, the remaining material may have a half-life of three or four years. On the other hand, the well-decomposed humic materials in soil are very stable and may have a half-life of hundreds of years. The result of the high rate of decomposition of fresh residues is that a small and difficult to predict portion of the soil's organic matter is providing much of the nitrogen being mineralized. Although there have been attempts to chemically characterize and estimate the amount of easily decomposable organic-N in soils, they have not resulted in a usable predictor of available N. Nitrogen availability, however, can be estimated based on manure type, application rates, and the previous crop. Such estimates are really indicators of the amount of active organic matter in the soil.

The amount of mineralization that occurs in a given year is also dependent on the weather. Regardless of organic matter content, cool and very wet weather patterns will result in lower amounts of mineralized N than warmer and less moist situations.

In addition to the rate of mineralization of organic-N, the loss of mineral N is an important factor governing N availability for crops. It is of little use to plants if a high amount of N is mineralized only to be lost before crops can use it. The amount of denitrification or leaching will be controlled to a large extent by the weather and soil characteristics, factors that cannot be anticipated by just knowing soil organic matter levels.

Predicting Nitrogen Availability

There is one situation in which organic matter or a factor correlated with organic matter may be helpful in predicting nitrogen availability: that of a given region with soils of similar texture and drainage class. In these soils, organic matter levels are almost entirely a reflection of management practices such as rotations, manure applications, and tillage. In this situation, percent organic matter may be well correlated with differences in nitrogen availability among soils, although it will not account for climatic effects.

Estimating the Nitrogen-Supplying Power of Soil

Knowing only the amount of organic matter does not usually help in estimating soil's nitrogen-supplying abilities. This explains why most of the proposed nitrogen-availability soil tests have not worked—the factors measured were correlated with organic matter or the total organic-N fraction, not with the N available to plants. While in the semiarid parts of the Midwest the amount of nitrate found in deep soil samples has proven useful for predicting nitrogen availability for crops, it took until the mid-1980s to develop a practical and accurate soil test for N availability in the humid regions of the United States. This test, developed for field corn, measures the nitrogen actually present as nitrate in the early part of the growing season. The timing of the test is such that it is still possible to apply nitrogen if the soil is found deficient.

Summary

There are a number of mathematical approaches to describing soil organic matter dynamics. One of these is a simple model in which the soil is assumed to be at equilibrium and any gains of organic matter just balance losses. By using an equilibrium model it is possible to see that there are different combinations of factors that can lead to the same amount of soil organic matter. Some more complex models assume a variety of organic fractions, each mineralizing at different rates, as well as a number of possible fates for a particular atom of carbon.

The amount of organic matter in a soil is usually not a good indicator of the potential nitrogen supply for plants. Neither is the nitrogen content of a soil, mainly consisting of organic forms, a good indicator of nitrogen availability to plants. The rate of decomposition is one of the main factors that controls soil organic matter levels. This means that there may be high levels of organic matter and nitrogen because of a low rate of decomposition. But a low rate of decomposition also means that not as much nitrogen will be mineralized and made available to plants as the high organic matter level might lead you to believe. Soil tests for nitrate have proven more useful than organic matter tests in assessing nitrogen availability to plants.

Sources

Albrecht, W. A. 1938. Loss of soil organic matter and its restoration. In *Soil and man.* 1938 Yearbook of Agriculture. Washington, D.C.: U.S.D.A.

Campbell, C. A. 1978. Soil organic carbon, nitrogen and fertility. In *Soil organic matter,* ed. M. Schnitzer and S. U. Khan, pp. 173–371. Developments in Soil Science 8. Amsterdam: Elsevier Scientific Publishing Co.

Jenkinson, D. S. 1988. Soil organic matter and its dynamics. In *Russell's soil conditions and plant growth,* ed. A. Wild, pp. 564–607. New York: John Wiley & Sons.

Jenkinson, D. S., and J. H. Rayner. 1977. The turnover of soil organic matter in some of the Rothamsted classical experiments. *Soil Science* 123:298–305.

Magdoff, F. R. 1991. Field nitrogen dynamics: Implications for assessing N availability. *Communications in Soil Science and Plant Analysis* 22:1507–17.

14

• • •

The Chemistry of Soil Organic Matter

Humus is definitely not a [specific] chemical compound, as was formerly believed. In fact, it is a very complex mixture of organic substances, and the proportion, and probably also the nature of the compounds, differ for each soil.—E. Fippin, 1913

Soil organic matter is about 50% carbon (C), 39% oxygen (O), 5% hydrogen (H), 5% nitrogen, 0.5% phosphorus (P), and 0.5% sulfur (S). Chains of carbon, with each carbon atom linked to other carbons, form the "backbone" of organic molecules. These carbon chains, with varying amounts of attached oxygen, hydrogen, nitrogen, phosphorus, and sulfur, are the basis for both simple sugars and amino acids and more complicated molecules of long carbon chains or rings.

There are many different types of organic molecules in soil. Some are simple molecules that have come directly from plants or other living organisms. These relatively simple chemicals—sugars, amino acids, and cellulose—are readily available for many organisms to use. For this reason they don't stay in the soil for a very long time as "free" molecules. Other chemicals such as resins and waxes also come directly from plants, but are more difficult for soil organisms to break down.

There are also slightly more complicated molecules produced by organisms living in the soil. For example, most of the soil polysaccharides, repeating units of sugar-type molecules connected in longer chains, are produced by microorganisms as they decompose fresh residues. Polysaccharides are known to promote better soil structure, and recent research indicates that the heavier polysaccharide molecules may be more important in promoting aggregate stability and water infiltration than the lighter molecules.

The characteristics of the well-decomposed part of the organic matter, the humus, are very different from those of simple organic molecules. There has been much interest in the chemistry of humic materials in soils and the amount present. About 65% of the organic molecules, the nonliving part of organic matter, is humus. It is known to be an important buffer, slowing down changes in soil acidity and nutrient availability. When compared with grains of sand, the sizes of the various types of humic compounds are very small. Humus and clay particles are *colloids,* because they are so small they can stay in a water suspension for very long times. But compared to simple organic molecules, humic substances are large, with high molecular weights, and very complex. While much is known about their general chemical composition, the relative significance of the various types of humic materials to plant growth is still not established.

Chemistry and Formation of Humus

The chemical structure of humus building blocks closely resembles that of lignin, as shown in figure 14.1. Both are based on the six-member benzene carbon ring shown in figure 14.1a. Humus tends to have more organic acid, or carboxyl (−COOH), than lignin, as you

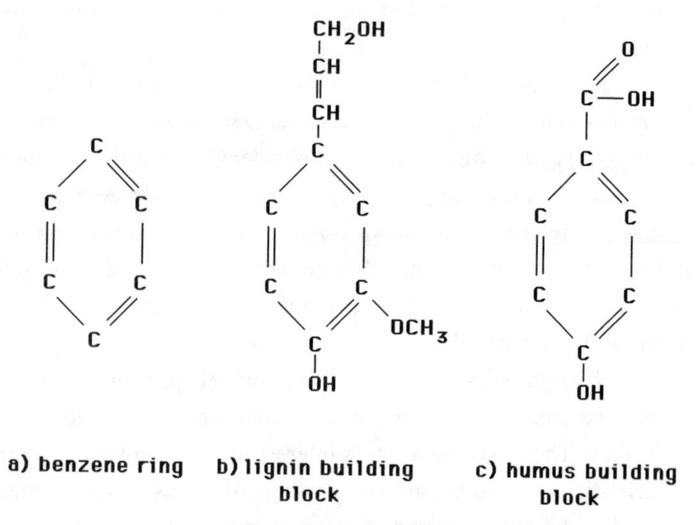

Figure 14.1. Structure of basic building blocks for lignin and humus.

can see at the top of figures 14.1b and c. The close resemblance between lignin and humus building blocks have led some scientists to believe that humus is formed when microorganisms modify lignin. However, it is also known that microorganisms are able to produce and secrete other molecules, such as phenolic compounds, that may be used to make humus. It is very possible that there are a number of ways that complex humic substances can be made in soils.

Characteristics of Humic Materials

It is not easy to get humus out of soils in the relatively pure form needed for analysis because it is intimately mixed with soil particles. Strong bases and acids are needed to accomplish this extraction. Humic materials are given different names depending on how they be-

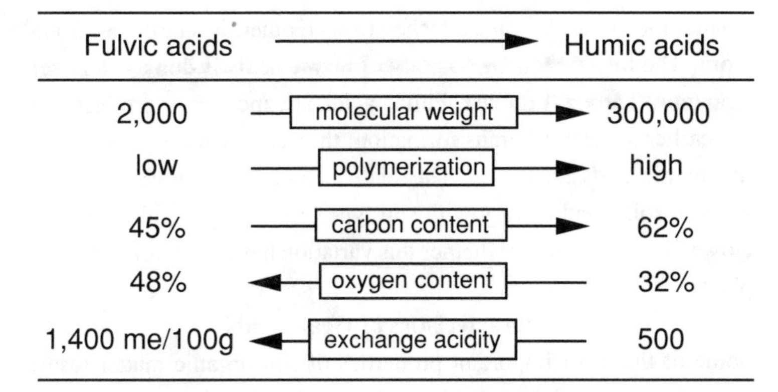

Figure 14.2. Chemical characteristics of fulvic and humic acids. Redrawn from F. J. Stevenson and J. H. A. Butler, Chemistry of humic acids and related substances, in *Organic Geochemistry,* ed. G. Englinton and M. J. T. Murphy (New York: Springer-Verlag, 1969), p. 536. Courtesy of Springer-Verlag.

have in base and acid. Humus that cannot be extracted with a strong base, such as sodium hydroxide (NaOH), is called *humin.* Humus extracted with a strong base but also soluble in strong acid is referred to as *fulvic acid.* The material that is extracted by base but is not soluble in acid is referred to as *humic acid.* The term *acid* is used to describe humic materials because humus behave like weak acids (see discussion below).

The humus extracted by a strong base is a complex mixture of large molecules. But there are large differences between the two main types of humus extracted from soils. *Exchange acidity* is the amount of acid that can be neutralized with a base. Exchange acidity is closely related to the *cation exchange capacity* (CEC) of organic matter. Compared with fulvic acids, humic acids tend to be larger and have a lower exchange acidity than fulvic acids (figure 14.2). Fulvic acids are converted to humic acids by a process called *polymerization,* in which smaller molecules become attached to each other by sharing a carbon atom. The lower exchange acidity of humic acids is due to its lesser amount of carboxyl groups. Fulvic acids are apparently produced in the earlier stages of humus formation; they can then be transformed into humic acids and humin. The relative amounts of humic and fulvic acids in soils tend to vary with soil type and management practices. However, it is not clear whether this variation has any significance for plants.

Characteristics of Weak Acids

Some of the most important properties of soil organic matter result from the nature of the organic acids associated with humus. Strong acids, such as nitric (HNO_3) and hydrochloric acid (HCl), are those that ionize readily, which means that most of their hydrogen is in solution as H^+. Weak acids, such as the organic acids in humus, do not give up their hydrogen easily. Hydrogen is part of the humus carboxyl (−COOH) under acidic conditions. When a soil is limed and the acidity decreases, there is a greater tendency for the H^+ to be removed from humic acids and to react with hydroxyl (OH^-) to form water. The carboxyl groups on the humus develop negative charge as the positively charged hydrogen is removed. When the pH of a soil is increased, the release of hydrogen from carboxyl groups helps to buffer the increase in pH and at the same time creates cation exchange capacity (negative charge). In the equation below, R represents the remaining part of the organic acid.

$$R{-}\overset{\overset{\displaystyle O}{/\!/}}{C}{-}OH + OH^- \rightarrow H_2O + R{-}\overset{\overset{\displaystyle O}{/\!/}}{C}{-}O^-$$

The formula describing the relationship of removal of the H from organic acids as pH changes is as follows:

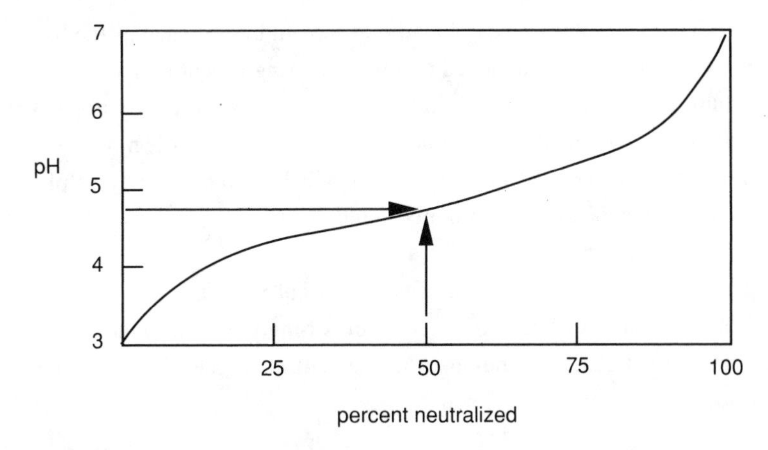

Figure 14.3. Relationship between the amount of acetic acid neutralized (H removed from carboxyl) and the pH. When half of the acid is neutralized, pH = pKa = 4.75.

$$pH = pKa + \log(R{-}COO^{-}/R{-}COOH) \qquad [1]$$

where R = the remaining part of the organic acid molecule, pKa = the pH of half neutralized acid, $RCOO^{-}$ = the amount of neutralized acid (moles/liter), and R−COOH = the amount of *un*neutralized acid (moles/liter).

When half of the acid's hydrogen is neutralized, $R{-}COO^{-}$ = R−COOH. The term $\log(R{-}COO^{-}/R{-}COOH)$ then becomes log(1). Because log(1) = 0, the pH at this point is equal to pKa. The pKa of carboxyl groups will vary depending on where they occur in the molecule and what kind of chemical groups are nearby.

In figure 14.3, the pH at the middle of the buffer curve, 4.75, is the point where 50% of the H^{+} is neutralized. This is the pKa of acetic acid. The pKa values for other simple organic acids are as follows: formic = 3.6; citric (with 3 carboxyl groups) = 3.1, 4.8, and 5.4; succinic (2 carboxyls) = 4.2 and 5.6; oxalic (2 carboxyls) = 1.3 and

4.3. There are other kinds of acidic groups in humus such as the hydroxyl that's directly attached to a benzene ring (phenol). The amine group, $R-NH_2$, is also an acid because it can also lose a hydrogen. But the pKa of phenol and amine groups in simple organic molecules indicates that most of their hydrogen will be removed only at pH values above the range commonly found in soils.

Weak Acid Characteristics of Humus

Both CEC and pH buffering of soils are strongly related to the weak-acid characteristics of humus. As we will soon see, they are like opposite sides of the same coin.

Cation Exchange Capacity (CEC)

The amount of negative charge that exists in a soil, referred to as cation exchange capacity, or CEC, is commonly expressed in terms of milliequivalents (1 me = 6×10^{20} charges) per 100 grams of soil. It can also be expressed in terms of centimoles (1 cmol = 6×10^{21} charges) per kilogram of soil. The CEC in a particular soil is due to the charges on both the clay minerals and the organic matter. The amount of charge on clay particles depends on the types of clays present. Clays composed mainly of aluminum and iron oxides, common in the southeastern United States and in the tropics, may have almost no CEC under acid conditions. In neutral pH soil these clays contribute about 0.03 me of charge for every one percent of clay in 100g. Whereas iron and aluminum oxide clays exhibit significant amounts of variable charge, the charges associated with most clays do not change greatly with pH. Kaolinitic clays, also common in the Southeast and in the tropics, have a low CEC and contribute about 0.1 me of charge to 100g of soil for every percentage point of clay. Illite, common in northeastern and midwestern soils, has an intermediate CEC and may contribute about 0.2 to 0.3 me of charge per percentage point of clay in 100g of soil. A soil that has about 20% illite will have a CEC of about 5 me for every 100 g of soil due to the presence of the clay (20 g of clay x 0.25 me per gram = 5 me). Montmorillonite, commonly found in high amounts in heavy clay soils derived from ocean-floor sediments or from basalt, has a relatively high CEC and may contribute about 1 me of charge for every percentage point of clay per 100g of soil.

Variable Charge of Soil Organic Matter

A consequence of the weak-acid nature of soil organic matter is that it develops a greater negative charge as the pH increases and hydrogen is released, leaving more and more negatively charged carboxyl groups. This effect, called *variable charge* or *pH-dependent CEC*, can be quite dramatic. For example, we found an increase in CEC after applying lime to a Calais loam with 5.6% organic matter. CEC went from 2.5 me/100 g soil at pH 4.2 to 14.0 me/100 g at pH 6.5.

After taking into account the clay content of our soil-testing-laboratory check samples, a relationship between the CEC of soil organic matter and pH was calculated:

$$\text{CEC (me/g soil om)} = -3.9 + 0.96(\text{pH}) \qquad [2]$$

The total soil CEC is the sum of the CEC on clays and the CEC of the organic matter. In soils that are low in clay content or contain low-CEC-type clays, the CEC of the organic matter may be a very high percentage of the total soil CEC. The strong influence that both organic matter and pH have on the CEC of soils can be seen in table 14.1. While the equation predicts no CEC on organic matter at pH 4, there will actually be a small amount of negative charge. For every one percent of organic matter in the soil, one unit of increase in pH results in a CEC increase of about 1 me per 100 g of soil.

Table 14.1. Influence of the Percentage of Organic Matter and pH on the Estimated CEC* due to Soil Organic Matter

	% Organic Matter					
	1	*2*	*3*	*4*	*5*	*6*
pH	CEC (me/100g soil)					
4	0	0	0	0	0	0
5	0.8	1.6	2.5	3.4	4.2	5.0
6	1.8	3.6	5.4	7.2	9.0	10.7
7	2.8	5.5	8.3	11.0	13.8	16.5

*Calculated by equation [2].

The equation used to calculate the values in table 14.1 predicts 0.4 and 12.7 me/100 g at pH 4.2 and 6.5 respectively for the sample of Calais loam soil previously discussed. These values compare fairly well to the 2.5 and 14.0 me/100g actually found in the experiment. One of the important results of liming acid soils is the increase in the ability of soil organic matter to hold cations because of an increase in CEC.

pH Buffering

The ability of soil organic matter to buffer a soil against pH changes is a result of the weak-acid phenomena discussed above. The organic acids are in an equilibrium relationship with the soil solution. Anything that changes one can also cause the other to change. As soil pH is increased by liming, hydrogen is released from organic-acid sites, neutralizing hydroxyl and carbonate ions in the solution. This is why it takes *tons* of limestone to increase the pH of a soil significantly compared to what would be needed to simply neutralize the free hydrogen present in the soil solution. All of the free hydrogen ions in the water of very acid soil (pH 4) could be neutralized with less than 5 pounds of limestone per acre. But from 2 to over 10 tons of limestone per acre will actually be needed to neutralize enough acidity in that soil to allow most crops to grow. Almost all of the hydrogen that must be neutralized when trying to increase soil pH is in organic acids, or associated with aluminum if the pH is very low.

Soil organic matter also helps buffer against a drop in pH by taking hydrogen from solution as the pH begins to drop.

$$R{-}C({=}O){-}O^- + H^+ \rightarrow R{-}C({=}O){-}OH$$

Many acid-forming reactions continually occur in soils. Some of these acids are produced as a result of organic matter decomposition by microorganisms, secretion by roots, or as oxidation reactions of inorganic elements. Commonly used nitrogen fertilizers are acid-forming in soils due to microbial conversion of NH_4^+ to NO_3^-. With high soil organic matter levels, however, the pH will decrease less rapidly and the field will have to be limed less frequently.

It was once thought that soil pH was strongly buffered in the range of 5 to 5.5, but it is now known that buffering is fairly uniform for most soils between pH 4.5 and 6.5. In addition, it is now realized that the greatest amount of buffering in most soils is caused by organic matter. Soil organic matter has many different types of organic acids each with a different pKa. Thus, it releases or takes up hydrogen from solution over a broad pH range.

Summary

Humus, the well-decomposed portion of soil organic matter, is a mixture of large, complex molecules. The composition of humus indicates that it may be derived from lignin, although there are most likely different ways for it to form in soil. Humus can partially counteract both physical and chemical changes that might cause soil fertility to deteriorate. This makes the soil more reliable for promoting plant growth.

Many of the important chemical properties of soil organic matter result from the weak-acid nature of humus. Cation exchange capacity (CEC) of organic matter, the ability to hold onto cations for plant use while protecting them from leaching, is due to the negative charges created as hydrogen is removed from weak acids during neutralization. When acids or bases are added to the soil, organic matter slows down, or buffers, the change in pH. Organic matter may provide nearly all of the CEC and pH buffering in soils low in clay or containing clays with little CEC.

Sources

Magdoff, F. R., and R. J. Bartlett. 1980. Effect of liming acid soils on potassium availability. *Soil Science* 129:12–14.

Magdoff, F. R., and R. J. Bartlett. 1985. Soil pH buffering revisited. *Soil Science Society of America Journal* 49:145–48.

Roberson, E. B., S. Selig, and M. K. Firestone. 1991. Cover crop management of polysaccharide-mediated aggregation in orchard soil. *Soil Science Society of America Journal* 55:734–39.

Schnitzer, M. 1991. Soil organic matter—the next 75 years. *Soil Science* 151:41–58.

Schnitzer, M., and S. U. Kahn, eds. 1978. *Soil organic matter.* Developments in Soil Science 8. Amsterdam: Elsevier Scientific Publishing Co.

Stevenson, F. J. 1982. *Humus chemistry: Genesis, composition, reactions*. New York: John Wiley & Sons.

Stevenson, F. J., and R. A. Olsen. 1989. A simplified representation of the chemical nature and reactions of humus. *Journal of Agronomic Education* 18:84–88.

• • •

Glossary

allelopathic effect. The effect that some plants have in suppressing the germination or growth of other plants. The chemicals responsible for this effect are produced during the growth of a plant or during the decomposition of its residues.

acid. A solution containing free hydrogen ions (H^+) or a chemical that will give off hydrogen ions into solution.

aggregates. The structures, or clumps, formed when soil minerals and organic matter are bound together with the help of organic molecules, plant roots, fungi, and clays.

anion. A negatively charged element or molecule such as chloride (Cl^-) or nitrate (NO_3^-).

ammonium (NH_4^+). A form of nitrogen that is available to plants and is produced in the early stage of organic matter decomposition.

available nutrient. The form of a nutrient that the plant is able to use. Nutrients that the plant needs are commonly found in the soil in forms that the plant can't use (such as organic forms of nitrogen) and must be converted into forms that the plant is able to take into the roots and use (such as the nitrate form of nitrogen).

base. Something that will neutralize an acid, such as hydroxide or limestone.

buffering. The slowdown or inhibition of changes or availability of nutrients in such things as soil acidity. A substance that has the ability to buffer a solution is also called a buffer. Buffering can slow down pH changes by neutralizing acids or bases.

calcareous soil. A soil in which finely divided lime is naturally distributed, usually have a pH of from 7 to slightly over 8.

cation. A positively charged ion such as calcium (Ca^{++}) or ammonium (NH_4^+).

cation exchange capacity (CEC). The amount of negative charge that exists on humus and clays, allowing them to hold onto positively charged chemicals (cations).

chelate. A molecule that uses more than one bond to attach strongly to certain elements such as iron (Fe^{++}) and zinc (Zn^{++}). These elements may later be released from the chelate and used by plants.

colloid. A very small particle, with a high surface area, that can stay in a water suspension for a very long time. The colloids in soils, the clay and humus molecules, are usually found in larger aggregates and not as individual particles. These colloids are responsible for many of the chemical and physical properties of soils, including cation exchange capacity, chelation of micronutrients, and the development of aggregates.

compost. Organic material that has been well decomposed by organisms under conditions of good aeration and high temperature, often used as a soil amendment.

conventional tillage. Preparation of soil for planting by using the moldboard plow followed by disking. It usually breaks down aggregates, buries most crop residues and manures, and leaves the soil smooth.

cover crop. A crop grown for the purpose of protecting the soil from erosion during the time of the year when the soil would otherwise be bare. It is sometimes also called a green manure crop.

C:N ratio. The amount of carbon in a residue divided by the amount of nitrogen. A high ratio results in low rates of decomposition and can also result in a temporary decrease in nitrogen nutrition for plants, as micro-organisms use much of the available nitrogen.

disk. An implement for harrowing, or breaking up, the soil. It is commonly used following a moldboard plow, but is also used by itself to break down aggregates, help mix fertilizers and manures with the soil, and smooth the soil surface.

element. All matter is made up of elements, seventeen of which are essential for plant growth. Elements such as carbon, oxygen, and nitrogen combine to form larger molecules.

green manure. A crop grown for the main purpose of building up or maintaining soil organic matter, it is also sometimes called a cover crop.

heavy soil. Soil that contains a lot of clay and is usually more difficult to work than lighter soil. It normally drains slowly following rain.

humus. The very well decomposed part of the soil organic matter. It has a high cation exchange capacity.

inorganic. Chemicals that are *not* made from chains or rings of carbon atoms (for example, soil clay minerals, nitrate, and calcium).

legume. A group of plants including beans, peas, clovers, and alfalfa that forms a symbiotic relationship with nitrogen-fixing bacteria living in their roots. These bacteria help to supply plants with an available source of nitrogen.

light soil. Sandy soil that can be worked easily and usually drains rapidly following rain.

lignin. A substance found in woody tissue and in stems of plants that is difficult for soil organisms to decompose.

lime, or limestone. A mineral that can neutralize acids and is commonly applied to acid soils, consisting of calcium carbonate ($CaCO_3$).

loess soils. Soils formed from windblown deposits of silty and fine-sand-size minerals. They are easily eroded by wind and water.

micronutrient. An element needed by plants in only small amounts, such as zinc, iron, copper, and manganese.

microorganism. Very small and simple organisms such as bacteria and fungi.

mineralization. The process by which soil organisms change organic elements into the "mineral" or inorganic form as they decompose organic matter (for example, organic forms of nitrogen are converted to nitrate).

moldboard plow. A commonly used plow that completely turns over the soil and incorporates any surface residues, manures, or fertilizers deeper into the soil.

monoculture. Production of the same crop in the same field year after year.

mycorrhizal relationship. The mutually beneficial relationship that develops between plant roots of most crops and fungi. The fungi help plants obtain water and phosphorus by acting like an extension of the root system and in return receive energy-containing chemical nutrients from the plant.

nitrate (NO_3^-). The form of nitrogen that is most readily available to plants. It is the nitrogen form normally found in the greatest abundance in agricultural soils.

nitrogen fixation. The conversion of atmospheric nitrogen by bacteria to a form that plants can use. A small number of bacteria, which include the rhizobia living in the roots of legumes, are able to make this conversion.

nitrogen immobilization. The transformation of available forms of nitrogen, such as nitrate and ammonium, into organic forms that are not readily available to plants.

no-till. A system of planting crops without tilling the soil with a plow, disk, chisel, or other implement.

organic. Chemicals that contain chains or rings of carbon connected to one another. Most of the chemicals in plants, animals, microorganisms, and soil organic matter are organic.

oxidation. The combining of a chemical such as carbon with oxygen, usually resulting in the release of energy.

pH. A way of expressing the acid status, or hydrogen ion (H^+) concentration, of a soil or a solution on a scale where 7 is neutral, less than 7 is acid, and greater than 7 is basic.

photosynthesis. The process by which green plants capture the energy of sunlight and use carbon dioxide from the atmosphere to make molecules needed for growth and development.

polyculture. Growth of more than one crop in a field at the same time.

polymerization. The process whereby small molecules of the same type attach to one another and form larger molecules made of repeating units.

respiration. The biological process that allows living things to use the energy stored in organic chemicals. In this process, carbon dioxide is released as energy is made available to do all sorts of work.

rhizobia bacteria. Bacteria that live in the roots of legumes and have a mutually beneficial relationship with the plant. These bacteria fix nitrogen, providing it to the plant in an available form, and in return receive energy-rich molecules that the plant produces.

rotation effect. The crop-yield benefit from rotations that includes better nutrient availability, fewer pest problems, and better soil structure.

silage. A feed produced when chopped-up corn plants or wilted hay are put into air-tight storage facilities (silos) and partially fermented by bacteria. The acidity produced by the fermentation and the lack of oxygen help preserve the quality of the feed during storage.

slurry (manure). A manure that is between solid and liquid. It flows slowly and has the consistency of something like a very thick soup.

sod crops. Grasses or legumes such as timothy and white clover that tend to grow very close together and form a fairly dense cover over the entire soil surface.

structure. The physical condition of the soil. It depends upon the amount of pores, the arrangement of soil solids into aggregates, and the degree of compaction.

texture. A name that indicates the relative significance of a soil's sand, silt, and clay content. The term "light texture" means that a soil has a high sand content, while "heavy texture" means that a soil has a high clay content.

thermophilic bacteria. Bacteria that live and work best under high temperatures, around 110° to 140°F. They are responsible for the most intense stage of decomposition that occurs during composting.

tilth. The physical condition, or structure, of the soil as it influences plant growth. A soil with good tilth is very porous and allows rainfall to infiltrate easily, permits roots to grow without obstruction, and is easy to work.

• • •

Sources for Further Information

ATTRA (Appropriate Technology Transfer for Rural Areas) is an organization funded by the U.S. Department of the Interior. It has developed a number of excellent informational packets on a wide variety of subjects including sustainable fertility management, cover crops and green manures, nutrient cycling in soils, and soil biology. ATTRA has a toll-free number, (800) 346-9140, and can also be reached at P.O. Box 3657, Fayetteville, Arkansas 72702.

The Alternative Farming Systems Information Center of the USDA National Agricultural Library has a Quick Bibliography Series with references such as *Soil Organic Matter: Impacts on Crop Production* (QB 91-24), *Composts and Composting of Organic Wastes* (QB 91-27), and *Manure: Uses, Costs and Benefits* (QB 90-71). For details about these and other bibliographies as well as other information contact the AFSIC at the National Agricultural Library, 10301 Baltimore Blvd., Beltsville, Maryland 20705-2351. Telephone: (301) 504-6559.

The New Farm is a magazine that has a lot of informative articles about the subjects covered in this book. The publisher's address is 222 Main Street, Emmaus, Pennsylvania 18098. Telephone: (215) 967-5171.

Most state Agricultural Extension Services have leaflets and booklets on manures and soil fertility and possibly green manures and other subjects dealt with in this book. You can request a list of publications from your Extension Service. Many states now have sustainable agriculture centers that publish newsletters. For example the California Sustainable Agriculture Research and Education Program publishes *Sustainable Agriculture News*. Information about this newsletter and other available materials can be obtained from UC SAREP, University of California, Davis, California 95616.

Soils, Soil Organisms, and Composting

Alexander, M. 1977. *Introduction to soil microbiology.* 2d ed. New York: John Wiley & Sons.

Brady, N. C. 1990. *The nature and properties of soils.* 10th ed. New York: Macmillan Publishing Co.

Dindal, D. n.d. *Ecology of compost.* Office of News and Publications, 122 Bray Hall, SUNY College of Environmental Science and Forestry, 1 Forestry Drive, Syracuse, New York 13210-2778. Telephone: (315) 470-6644.

Martin, D. L., and G. Gershuny, eds. 1992. *The Rodale book of composting: Easy methods for every gardener.* Emmaus, Penn.: Rodale Press.

Paul, E. A., and F. E. Clark. 1989. *Soil microbiology and biochemistry.* San Diego: Academic Press.

Rynk, R., ed. 1992. *On-farm composting.* NRAES-54. Ithaca, N.Y.: Northeast Regional Agricultural Engineering Service. This very useful book can be ordered from NRAES, 152 Riley Robb Hall, Cooperative Extension, Ithaca, NY 14853-5701.

Werner, M. R., and D. L. Dindal. 1990. Effects of conversion to organic agricultural practices on soil biota. *American Journal of Alternative Agriculture* 5(1):24–32.

Manures, Fertilizers, Tillage, and Rotations

Cramer, Craig, and the Editors of *The New Farm,* eds. 1986. *The farmer's fertilizer handbook.* Emmaus, Penn.: American Regenerative Agriculture Association. This handbook contains lots of very good information on soil fertility, soil testing, use of manures, and use of fertilizers.

Elliott, L. F., and F. J. Stevenson, eds. 1977. *Soils for management of organic wastes and wastewaters.* Madison, Wis.: Soil Science Society of America.

Follett, R. F., J. W. B. Stewart, and C. V. Cole, eds. 1987. *Soil fertility and organic matter as critical components of production systems.* Madison, Wis.: American Society of Agronomy.

Francis, C. A., and M. D. Clegg. 1990. Crop rotations in sustainable production systems. In *Sustainable agricultural systems,* ed. C. A. Edwards, R. Lal, P. Madden, R. H. Miller, and G. House. Ankeney, Iowa: Soil and Water Conservation Society.

Lal, R., and F. J. Pierce, eds. 1991. *Soil management for sustainability.* Ankeny, Iowa: Soil and Water Conservation Society.

Parnes, R. 1990. *Fertile soil—a growers guide to organic and inorganic fertilizers.* Davis, California: agAccess.

Cover Crops

Hargrove, W. L., ed. 1991. *Cover crops for clean water.* Ankeny, Iowa: Soil and Water Conservation Society.

Pieters, A. J. 1927. *Green manuring principles and practices.* New York: John Wiley & Sons. An oldie but goody—you'll probably have to go to the library of a College of Agriculture to find this.

Power, J. F., ed. 1987. *The role of legumes in conservation tillage systems.* Ankeny, Iowa: Soil Conservation Society of America.

Rodale Institute. 1992. *Managing cover crops profitably.* Sustainable Agriculture Network, Handbook Series #1. USDA Sustainable

Agriculture Research and Education Program. This book was written especially for farmers and agricultural agents and includes the latest information in a very usable form. It can be ordered for $9.95 from Sustainable Agriculture Publications, Hills Building, University of Vermont, Burlington, VT 05405.

Dynamics and Chemistry of Organic Matter

Jenkinson, D. S. 1988. *Soil organic matter and its dynamics*. In Russell's soil conditions and plant growth, ed. A. Wild, pp. 564–607. New York: John Wiley & Sons.

Lucas, R. E., J. B. Holtman, and J. L. Connor. 1977. Soil carbon dynamics and cropping practices. In *Agriculture and energy,* ed. W. Lockeretz, pp. 333–51. New York: Academic Press.

Schnitzer, M., and S. U. Kahn, eds. 1978. *Soil organic matter.* Developments in Soil Science 8. Amsterdam: Elsevier Scientific Publishing Co.

Stevenson, F. J. 1982. *Humus chemistry: Genesis, composition, reactions*. New York: John Wiley & Sons.

• • •

Index